Collins

Collins Revision

GCSE Higher Chemistry

Revision Guide

FOR OCR GATEWAY B

About this book

This book covers GCSE Chemistry for OCR Gateway B at Higher Level. Written by GCSE examiners, it is designed to help you to get the best grade in your GCSE Chemistry exams.

The book is divided into three parts; a topic-by-topic revision guide, workbook practice pages for each topic and detachable answers.

How to Use It

The revision guide section gives you complete coverage of each of the six modules that you need to study. Use it to build your knowledge and understanding.

The workbook section is packed with exam-style questions. Once you have covered a particular topic, use the matching workbook page to test yourself.

The answers in the back of the book are detachable. Remove them to help check your grade or a friend's.

Go Up a Grade

There are lots of revision guides for you to choose from. This one is different because it really helps you to go up a grade. Each topic in the revision guide and workbook sections is broken down and graded to show you what examiners look for at each level. This lets you check where you are, and see exactly what you need to do to improve your grade at every step. Crucially, it shows you what makes the difference between an D–C and a B–A* grade answer.

Special Features

- **Questions** at the end of every topic page quickly test your level.
- **Top Tips** give you extra advice about what examiners really want.
- **Summaries** of each module remind you of the most important things to remember.
- **Checklists** for each module help you to monitor your progress.
- A comprehensive **glossary** gives you a quick reference guide to the Chemistry terms that you need to know.

Published by Collins
An imprint of HarperCollins*Publishers*
77–85 Fulham Palace Road
Hammersmith
London W6 8JB

Browse the complete Collins catalogue at
www.collinseducation.co.uk

© HarperCollins*Publishers* Limited 2010

10 9 8 7 6 5 4 3

ISBN-13 978-0-00-734809-1

The authors assert their moral rights to be identified as the authors of this work.

All rights reserved. No part of this publication may be reproduced, stored in a retrieval system, or transmitted in any form or by any means, electronic, mechanical, photocopying, recording or otherwise, without the prior written permission of the Publisher or a licence permitting restricted copying in the United Kingdom issued by the Copyright Licensing Agency Ltd., 90 Tottenham Court Road, London W1T 4LP.

British Library Cataloguing in Publication Data
A Catalogue record for this publication is available from the British Library

Written by Ann Daniels
Series Consultant Chris Sherry
Project Manager Charis Evans
Design and layout Graham Brasnett
Editor Mitch Fitton
Illustrated by Kathy Baxendale, IFA design Ltd, Mark Walker, Bob Lea and Steve Evans
Indexed by Marie Lorimer
Printed and bound in China by
South China Printing Company

Acknowledgements

The Authors and Publishers are grateful to the following for permission to reproduce photographs:

Photos.com p7, istock photo p10, Charles D Winters/Science Photo Library p28, Andrew Lambert Photography/Science Photo Library p29, Dr Jeremy Burgess/Science Photo Library p39, istock p59.

Whilst every effort has been made to trace the copyright holders, in cases where this has been unsuccessful, or if any have inadvertently been overlooked, the Publishers will be pleased to make the necessary arrangements at the first opportunity.

Contents

		Revision	Workbook
C1 Carbon chemistry	Cooking	5	63
	Food additives	6	64
	Smells	7	65
	Making crude oil useful	8	66
	Making polymers	9	67
	Designer polymers	10	68
	Using carbon fuels	11	69
	Energy	12	70
	C1 Summary	13	71

		Revision	Workbook
C2 Rocks and metals	Paints and pigments	14	72
	Construction materials	15	73
	Does the Earth move?	16	74
	Metals and alloys	17	75
	Cars for scrap	18	76
	Clean air	19	77
	Faster or slower (1)	20	78
	Faster or slower (2)	21	79
	C2 Summary	22	80

		Revision	Workbook
C3 The periodic table	What are atoms like?	23	81
	Ionic bonding	24	82
	Covalent bonding	25	83
	The group 1 elements	26	84
	The group 7 elements	27	85
	Electrolysis	28	86
	Transition elements	29	87
	Metal structure and properties	30	88
	C3 Summary	31	89

		Revision	Workbook
C4 Chemical economics	Acids and bases	32	90
	Reacting masses	33	91
	Fertilisers and crop yield	34	92
	The Haber process	35	93
	Detergents	36	94
	Batch or continuous?	37	95
	Nanochemistry	38	96
	How pure is our water?	39	97
	C4 Summary	40	98

		Revision	Workbook
C5 How much?	Moles and empirical formulae	41	99
	Electrolysis	42	100
	Quantitative analysis	43	101
	Titrations	44	102
	Gas volumes	45	103
	Equilibria	46	104
	Strong and weak acids	47	105
	Ionic equations	48	106
	C5 Summary	49	107

		Revision	Workbook
C6 Chemistry out there	Energy transfers – fuel cells	50	108
	Redox reactions	51	109
	Alcohols	52	110
	Chemistry of sodium chloride (NaCl)	53	111
	Depletion of the ozone layer	54	112
	Hardness of water	55	113
	Natural fats and oils	56	114
	Analgesics	57	115
	C6 Summary	58	116

The periodic table	4
How Science Works	59–60
Fundamental concepts	62
Glossary	117–20
Answers	121–8

1	2											3	4	5	6	7	8
						1 **H** hydrogen 1											4 **He** helium 2
7 **Li** lithium 3	9 **Be** beryllium 4											11 **B** boron 5	12 **C** carbon 6	14 **N** nitrogen 7	16 **O** oxygen 8	19 **F** fluorine 9	20 **Ne** neon 10
23 **Na** sodium 11	24 **Mg** magnesium 12											27 **Al** aluminium 13	28 **Si** silicon 14	31 **P** phosphorus 15	32 **S** sulfur 16	35.5 **Cl** chlorine 17	40 **Ar** argon 18
39 **K** potassium 19	40 **Ca** calcium 20	45 **Sc** scandium 21	48 **Ti** titanium 22	51 **V** vanadium 23	52 **Cr** chromium 24	55 **Mn** manganese 25	56 **Fe** iron 26	59 **Co** cobalt 27	59 **Ni** nickel 28	63.5 **Cu** copper 29	65 **Zn** zinc 30	70 **Ga** gallium 31	73 **Ge** germanium 32	75 **As** arsenic 33	79 **Se** selenium 34	80 **Br** bromine 35	84 **Kr** krypton 36
85 **Rb** rubidium 37	88 **Sr** strontium 38	89 **Y** yttrium 39	91 **Zr** zirconium 40	93 **Nb** niobium 41	96 **Mo** molybdenum 42	[98] **Tc** technetium 43	101 **Ru** ruthenium 44	103 **Rh** rhodium 45	106 **Pd** palladium 46	108 **Ag** silver 47	112 **Cd** cadmium 48	115 **In** indium 49	119 **Sn** tin 50	122 **Sb** antimony 51	128 **Te** tellurium 52	127 **I** iodine 53	131 **Xe** xenon 54
133 **Cs** caesium 55	137 **Ba** barium 56	139 **La*** lanthanum 57	178 **Hf** hafnium 72	181 **Ta** tantalum 73	184 **W** tungsten 74	186 **Re** rhenium 75	190 **Os** osmium 76	192 **Ir** iridium 77	195 **Pt** platinum 78	197 **Au** gold 79	201 **Hg** mercury 80	204 **Tl** thallium 81	207 **Pb** lead 82	209 **Bi** bismuth 83	[209] **Po** polonium 84	[210] **At** astatine 85	[222] **Rn** radon 86
[223] **Fr** francium 87	[226] **Ra** radium 88	[227] **Ac*** actinium 89	[261] **Rf** rutherfordium 104	[262] **Db** dubnium 105	[266] **Sg** seaborgium 106	[264] **Bh** bohrium 107	[277] **Hs** hassium 108	[268] **Mt** meitnerium 109	[271] **Ds** darmstadtium 110	[272] **Rg** roentgenium 111							

Key:
relative atomic mass
atomic symbol
name
atomic (proton) number

Elements with atomic numbers 112–116 have been reported but not fully authenticated.

* The Lanthanides (atomic numbers 58–71) and the Actinides (atomic numbers 90–103) have been omitted.
Cu and Cl have not been rounded to the nearest whole number.

Cooking

Cooking food

Grades D–C

- Some foods can be eaten **raw**, but other foods must be **cooked** to make them safer or more attractive. Food is cooked because:
  - the high temperature **kills** harmful **microbes** in food
  - the **texture** of food is improved
  - the **taste** of food is improved
  - the **flavour** of food is enhanced
  - cooked food is easier to **digest**.

Proteins

Grades D–C

- Potatoes and flour are good sources of **carbohydrate**.
- Meat and eggs are good sources of **proteins**. Proteins are large molecules that have definite shapes. When food is cooked, the protein molecules change shape.

Proteins denature on heating.

Grades B–A*

- The shape change is **irreversible** and the protein molecule is now **denatured**.

Baking powder

Grades D–C

- Baking powder is a chemical called **sodium hydrogencarbonate**.
- When it's heated, it **decomposes** to give sodium carbonate, carbon dioxide and water:
  - the **reactant** is sodium hydrogencarbonate
  - the **products** are sodium carbonate, carbon dioxide and water.
- The word equation for the reaction is:

  sodium hydrogencarbonate $\xrightarrow{heat}$ sodium carbonate + carbon dioxide + water

- It's the **carbon dioxide** in the reaction that helps **cakes** to **rise**.

Grades B–A*

- The balanced symbol equation for this is:

  $2NaHCO_3 \longrightarrow Na_2CO_3 + H_2O + CO_2$

Testing for carbon dioxide

Grades D–C

- The chemical test for **carbon dioxide** is to pass it through limewater. It turns the **limewater** from colourless to milky white.

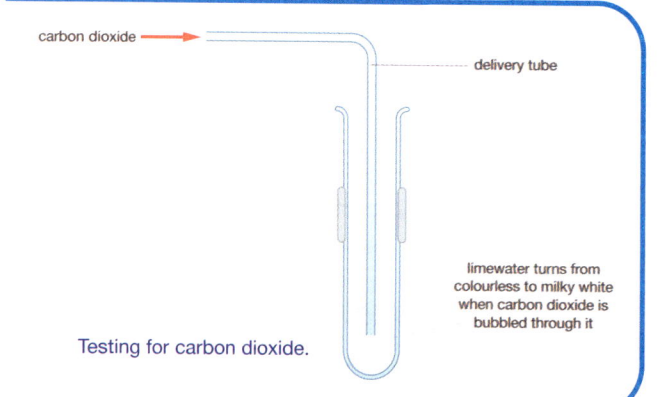

Testing for carbon dioxide.

Questions

Grades D–C
1 What do protein molecules do when they're heated?

Grades B–A*
2 When heated, proteins 'denature'. What does this mean?

Grades D–C
3 Jo says that when you heat baking powder, it decomposes and three products are made. Explain what she means.

Grades B–A*
4 Write a symbol equation for the decomposition of baking powder.

C1 CARBON CHEMISTRY

Food additives

Food additives

- Food additives are added for different reasons:
  – to **preserve** food from reacting with oxygen, bacteria or mould
  – to give a different **sensory experience**, such as to **enhance** the colour or flavour of food.
- **Antioxidants** stop food from reacting with oxygen. **Ascorbic acid** (vitamin C) is used in tinned fruit and wine as an antioxidant. Its E number is E300.
- Information about food is given on food **labels**.

Food packaging

- **Intelligent** or **active packaging** are methods used to stop food spoiling. They remove water or heat or cool the contents of packs.
  – **Active packaging** changes the condition of the food to extend its shelf life.
  – **Intelligent packaging** uses sensors to monitor the quality of the food and lets the customer know when the food is no longer fresh.

- **Active packaging** uses a **polymer** and a **catalyst** as a packaging film that scavenges for oxygen. It prevents the need for additives, such as antioxidants, to be added to foods. It often involves the removal of water to make bacteria or mould more difficult to grow. It's used for cheese and fruit juice.

- **Intelligent packaging** includes **indicators** on packages. An indicator on the outside of a package shows how fresh a food is. A central circle darkens as the product loses its freshness.

An indicator on intelligent packaging.

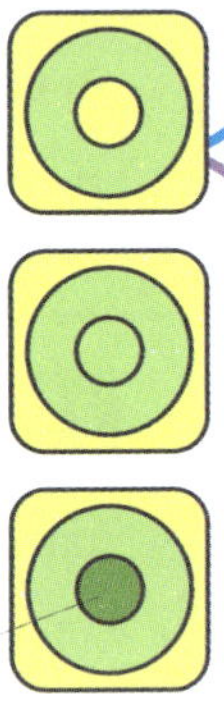

fresh

still fresh – consume immediately

no longer fresh

central circle darkens as food loses its freshness

Emulsions and emulsifiers

- Detergents are long molecules made up of two parts: a head and a tail. The tail is a 'fat-loving' part and the head is a 'water-loving' part.

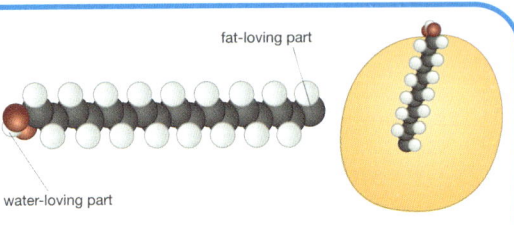

- Examples of **emulsions** are:
  – some **paints**
  – **milk**, which is an emulsion of oil in water
  – **mayonnaise**, which is an emulsion of oil and vinegar with egg. Egg is the **emulsifier**.

fat-loving part

water-loving part

The fat-loving part of the detergent goes into the grease droplet.

- When mayonnaise is made, the egg yolk binds the oil and vinegar together to make a smooth substance.

- The mayonnaise doesn't separate as the egg yolk has a molecule that has two parts:
  – a water-loving part that attracts vinegar to it, called the **hydrophilic head**
  – a water-hating part that attracts oil to it, called the **hydrophobic tail**.

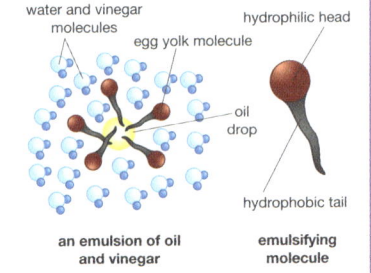

water and vinegar molecules
egg yolk molecule
hydrophilic head
oil drop
hydrophobic tail

an emulsion of oil and vinegar

emulsifying molecule

An emulsifying molecule.

- The hydrophobic tail is attracted into the lump of oil but the head isn't. The hydrophilic head is attracted to water and 'pulls' the oil on the tail into the water.

Questions

(Grades D–C)
1 What's the antioxidant added to tinned fruit and wine?

(Grades B–A*)
2 How does the removal of water help to preserve food?

(Grades D–C)
3 Emulsifiers are long molecules that have two parts. Describe the two parts.

(Grades B–A*)
4 Emulsifiers keep oil and water from separating. Explain how.

Smells

Esters

- To make a perfume, **alcohol** is mixed with an **acid** to make an **ester**.

  alcohol + acid ⟶ ester + water

- Look at the diagram on the right.
  - Acid is added to the alcohol and is heated for some time.
  - The condenser stops the gas from escaping and helps it to cool down again.

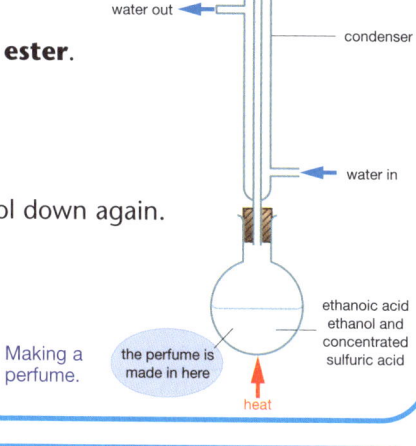

Making a perfume.

water out, condenser, water in, ethanoic acid ethanol and concentrated sulfuric acid, the perfume is made in here, heat

Perfume properties

- A **perfume** needs to:
  - **evaporate** easily so the perfume particles can reach the nose
  - be **non-toxic** so it doesn't poison people
  - be **insoluble** in water so it can't be washed off easily
  - not react with water so it doesn't react with perspiration
  - not irritate the skin so it can be sprayed directly onto the skin.

Solutions

- A **solution** is a solute and a solvent that don't separate.

- Esters are used as **solvents**. Other solvents can be used as cleaners, such as oil to clean grease.

- **Cosmetics** need to be thoroughly **tested** so that they don't harm humans.
  - They shouldn't cause rashes or itchiness.
  - They shouldn't cause skin damage or lead to cancer or other life-threatening conditions.

- Cosmetic testing takes many years and is highly controversial.
  - Some people object to testing on animals as the animals may be harmed, and don't have any control over what happens to them.
  - Other people say they feel safer if the cosmetics have been tested on animals.

Particles

- If a liquid evaporates easily then the substance is **volatile**.

- **Particles** in liquid **perfume** are weakly attracted to each other. When some of these particles increase their **kinetic energy**, the force of attraction between them is overcome. The particles escape through the surface of the liquid into the surroundings as gas particles. This is **evaporation**. The gas particles move through the air by **diffusion** to reach the sensors in the nose.

- Water doesn't dissolve **nail varnish**. This is because the force of attraction between two water molecules is stronger than that between a water molecule and a molecule of nail varnish. Also, the force of attraction between two nail varnish molecules is stronger than between a nail varnish molecule and a water molecule.

Questions

Grades D–C
1. Which two substances combine to make an ester?
2. Two people may have different ideas about cosmetic testing. Explain why.

Grades B–A*
3. What does 'volatile' mean?
4. Louise removes her nail varnish with a solvent. Explain why she doesn't use water.

Making crude oil useful

Fossil fuels

- **Fossil fuels** are **finite resources** because they are no longer being made. When these fossil fuels are used up, there will be no more. They're called a **non-renewable** source.

Fractional distillation

- **Crude oil** is a mixture of many types of oil, which are all **hydrocarbons**.
- In **fractional distillation**, crude oil is heated at the bottom of a tower.
  - Oil that doesn't boil, sinks as a thick liquid to the bottom. This fraction is **bitumen** and is used to make **tar** for road surfaces. Bitumen has a very high boiling point. It 'exits' at the bottom of the tower.
  - Other fractions boil and their gases rise up the tower. Fractions with lower boiling points, such as **petrol** and **LPG**, 'exit' at the top of the tower, where it's colder.

A fractional distillation column.

- Crude oil can be separated because the molecules in different fractions have **different length chains**. The **forces** between the molecules (**intermolecular forces**) are different and are broken during boiling. The molecules of a liquid separate from each other as molecules of gas.
  - **Heavy molecules**, such as in bitumen and heavy oil, have very long chains, so the molecules have strong forces of attraction. This means they're difficult to separate. A lot of energy is needed to pull the molecules apart. They have **high boiling points**.
  - **Lighter molecules**, such as petrol, have short chains. Each molecule has weak attractive forces and is easily separated. Less energy is needed to pull them apart. They have very **low boiling points**.

Problems with extracting crude oil

- **Oil slicks** can harm animals, pollute beaches and destroy unique habitats. Clean-up operations are extremely expensive and the detergents and barrages used can cause **environmental problems**.

- Extracting crude oil can cause **political problems**. Oil-producing nations can set prices high and cause problems for non-oil producing nations.

Cracking

Cracking.

- Cracking is a process that:
  - turns large alkane molecules into smaller alkane and alkene molecules
  - also makes useful alkene molecules with a **double bond**, which can be used to make **polymers**.
- Alkanes have a general formula of: C_nH_{2n+2}.

Octane has 8 carbon atoms and $2n + 2 = 18$ hydrogen atoms. The formula for octane is C_8H_{18}.

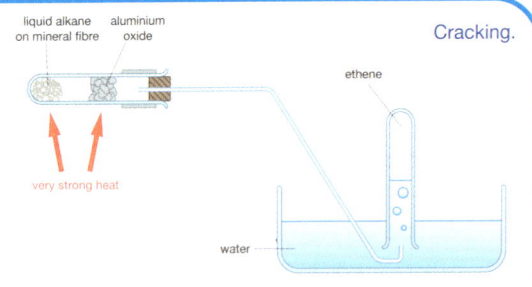

Questions

(Grades D–C)
1 Where do fractions with the lowest boiling points 'exit' the tower in fractional distillation?

(Grades B–A*)
2 Which products from crude oil have strong forces of attraction between the molecules? Explain why.

(Grades D–C)
3 What's the formula for an alkane with 7 carbon atoms?

(Grades B–A*)
4 How does industry match the demand for petrol with the supply from crude oil?

Making polymers

Polymerisation

Grades D–C

- **Polymerisation** is the process in which many **monomers** react to give a **polymer**. This reaction needs **high pressure** and a **catalyst**.

- You can recognise a polymer from its **displayed formula** by looking out for these features:
  – a long chain
  – the pattern repeats every two carbon atoms
  – there are two brackets on the end with extended bonds through them
  – there's an 'n' after the brackets.

  This is the displayed formula of poly(ethene).

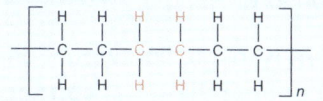

Grades B–A*

- **Addition polymerisation** is the reaction of many monomers of the same type that have **double bonds** to form a polymer that has **single bonds**.

- The displayed formula of an addition polymer can be constructed when the displayed formula of its monomer is given.

  The displayed formula for the ethene monomer is:
  During a polymerisation reaction, the high pressure and catalyst cause the double bond in the ethene monomer to break and each of the two carbon atoms forms a new bond. The reaction continues until it's stopped, making a long chain. This is poly(ethene).

- If the displayed formula of an addition polymer is known, the displayed formula of its monomer can be worked out by looking at its repeated units.

  This addition polymer: has a repeated unit of two carbon atoms, three hydrogen atoms and one chlorine atom. Therefore the monomer's displayed formula is:

- An **unsaturated** compound contains at least one double bond between carbon atoms. A **saturated** compound contains only single bonds between carbon atoms.

Hydrocarbons

Grades D–C

- A hydrocarbon is a compound of carbon and hydrogen atoms only.
  – An **alkane** has a single bond C–C.   – An **alkene** has one double bond C=C.

Propane, C_3H_8, is a hydrocarbon and an alkane.

Propanol, C_3H_7OH, isn't a hydrocarbon because it contains an oxygen atom.

Propene is a hydrocarbon, an alkene and a **monomer**.

Polypropene is the **polymer**.

Grades B–A*

- **Alkenes** are **unsaturated**. The general formula of an alkene is: C_nH_{2n}.
  Alkanes are **saturated**. The general formula of an alkane is: C_nH_{2n+2}. It has no double bonds.

- Carbon and hydrogen atoms share an electron pair to form **covalent bonds**.

- **Bromine solution** is used to test for **unsaturation**. When an alkene is added, the orange bromine solution turns colourless because it has reacted with the alkene to form a new compound. An alkane doesn't react with bromine solution and so the bromine remains orange.

Questions

Grades D–C
1 What two conditions are needed for polymerisation to take place?

Grades B–A*
2 Look at the polymer. Draw its monomer.

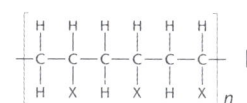

Grades D–C
3 What's the difference between an alkane and an alkene?

Grades B–A*
4 What's the general formula for an alkene?

Designer polymers

Breathable polymers

- Polymers are better than other materials for some uses.

use	polymer	other material
contact lens	wet on the eye	dry on the eye
teeth filling	attractive	looks metallic
wound dressing	waterproof	gets wet

Waterproof walking.

- **Nylon** is tough, lightweight and keeps rainwater out, but it keeps body sweat in. The water vapour from the sweat **condenses** and makes the wearer wet and cold inside their raincoat.

- If nylon is **laminated** with a PTFE/polyurethane **membrane**, clothing can be made that's **waterproof** and **breathable**. Gore-Tex® has all the properties of nylon and is breathable, so it's worn by many active outdoor people. Water vapour from sweat can pass through the membrane, but rainwater can't.

- In Gore-Tex materials, the inner layer of the clothing is made from expanded PTFE (polytetrafluoroethene), which is **hydrophobic**.

- The PTFE is expanded to form a **microporous membrane**. Only small amounts of the polymer are needed to create an airy, lattice-like structure. Wind doesn't pass through the membrane.

- In expanded PTFE a membrane pore is 700 times larger than a water vapour molecule and therefore moisture from sweat passes through.

Biodegradable and non-biodegradable polymers

- Scientists are developing **addition polymers** that are **biodegradable**. These are disposed of easily by **dissolving**. Biopol is a biodegradeable plastic that can be used to make laundry bags for hospitals. It degrades when washed leaving the laundry in the machine.

- **Disposal problems** for **non-biodegradable** polymers include the following:
  – landfill sites get filled quickly and waste valuable land
  – burning waste plastics produces toxic gases
  – disposal by burning or landfill sites wastes a valuable resource
  – problems in sorting different polymers makes recycling difficult.

Stretchy polymers and rigid polymers

- The **atoms** of the monomers in each of the chains in a polymer are held together by **strong intramolecular bonds**. The **chains** in the polymer are held together by weak intermolecular forces of attraction.
  – Plastics that have **weak intermolecular forces of attraction** between the polymer molecules have **low melting points** and can be **stretched** easily as the polymer molecules can slide over one another.
  – Other plastics that form strong **intermolecular chemical bonds** or cross-linking bridges between the polymer molecules have **high melting points** and can't be stretched easily as the molecules are rigid.

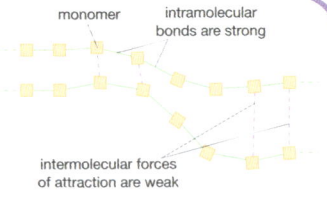

Intermolecular bonds are stronger than the intermolecular forces of attraction.

Questions

Grades D–C
1 What's the disadvantage of using nylon to make outdoor clothes?

Grades B–A*
2 How big are the pores in the layer in Gore-Tex® compared to the size of water vapour molecules?

Grades D–C
3 What's given off when disposing of plastics by burning?

Grades B–A*
4 Some polymers stretch easily. Explain why.

Using carbon fuels

Choosing fuels

- A fuel is chosen because of its **characteristics**.

characteristic	coal	petrol
energy value	high	high
availability	good	good
storage	bulky and dirty	volatile
toxicity	produces acid fumes	produces less acid fumes
pollution caused	acid rain, carbon dioxide and soot	carbon dioxide, nitrous oxides

- Coal produces pollution in the form of **sulfur dioxide**. This dissolves to produce **acid rain**, which damages stone buildings and statues, and kills fish and trees.

- Petrol and diesel are **liquids** so they can circulate easily in an engine, and can be stored in petrol stations along road networks.

- As fossil fuels are easy to use and the population is increasing, more are being consumed. More **carbon dioxide**, which is a **greenhouse gas**, is released and contributes to **climate change**. Many governments have pledged to try to cut carbon dioxide emissions over the next 15 years. It's a **global problem** that can't be solved by one country alone.

Combustion

- A **blue flame** means the fuel is burning in **plenty of oxygen**: **complete combustion**.

- A **yellow flame** means the fuel is burning in a **shortage of oxygen**: **incomplete combustion**.

- The word equation for a fuel burning in air is:
  fuel + oxygen ⟶ carbon dioxide + water

- This reaction can be shown by an experiment in the laboratory.
  – **Limewater** is used to test for carbon dioxide.
  – **White copper sulfate** powder is used to test for water, which is produced as steam.

- Complete combustion is better than incomplete combustion because: **less soot** is made, more **heat energy** is **released**, and toxic **carbon monoxide** gas isn't produced.

- A **heater** in a poorly ventilated room is burning fuel in a shortage of oxygen. It gives off poisonous carbon monoxide. Gas appliances should be checked regularly.

- Complete combustion releases useful energy. The formulae for the **products** of complete combustion are: carbon dioxide CO_2 and water H_2O.
  – **Methane** is a common hydrocarbon fuel. The formula for methane is CH_4.
  – The equation for complete combustion is: $CH_4 + O_2 \longrightarrow CO_2 + H_2O$
  The equation must be made to **balance**. There are two oxygen atoms in the reactants and three oxygen atoms in the products.
  The balanced equation is: $CH_4 + 2O_2 \longrightarrow CO_2 + 2H_2O$

When a fuel burns in air, what's produced?

Top Tip! First count up the numbers of atoms in each molecule (shown by the subscript numbers). Don't change these numbers. Then, if necessary, add to the molecule number (large number in front of formula) to balance the numbers on either side of the arrow.

Questions

Grades B–A*
1. Explain why governments are concerned about increased carbon dioxide emissions.

Grades D–C
2. Which liquid is used to test for carbon dioxide?

3. Complete combustion is better than incomplete combustion. Explain why.

Grades B–A*
4. Write a balanced equation for the complete combustion of pentane: C_5H_{12}.

Energy

Types of reaction

- Chemical reactions can be divided into two groups:

exothermic	endothermic
energy transferred to surroundings (energy is released)	energy taken from surroundings (absorbs energy)
temperature increases	temperature decreases
bonds made	bonds are broken
less energy needed to break bonds than make new bonds, reaction is exothermic overall	more energy needed to break bonds than make new bonds, reaction is endothermic overall

Comparing the energy from different flames

- The flame of a Bunsen burner changes colour depending on the amount of oxygen it burns in:
  – blue flames are seen when the gas burns in plenty of oxygen (**complete combustion**)
  – yellow flames are seen when the gas burns in limited oxygen (**incomplete combustion**).

- To design an experiment to compare the energy transferred in the two different flames, remember:
  – the apparatus used to compare fuels – the amount of gas used needs to be measured
  – make the tests fair.

Comparing fuels using calculations

- The amount of energy transferred during a reaction can be calculated.
  – A spirit burner or a bottled gas burner is used to heat water in a copper calorimeter.
  – A temperature change is chosen and measured, for example 50 °C.
  – The mass of fuel burnt is measured by finding the mass before and after burning.

- The **tests** are made **fair** by having the:
  – same mass of water – same temperature change
  – same distance from the calorimeter to the flame.

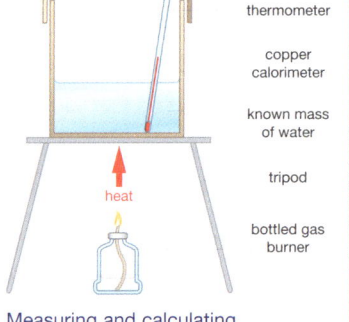

- The tests are made **reliable** by repeating the experiment three times and excluding draughts.

- To calculate the **energy transferred**, use the following formula. The unit is **joules** (J).
  energy transferred = mass of water × 4.2 × temperature change

  Calculate the energy transfer if 100 g of water is heated from 20 °C to 70 °C.
  energy transferred = 100 × 4.2 × (70 − 20) = 420 × 50 = 21 000 J (21 kJ)

- To calculate the **energy output** of a fuel, use the following formula. The unit is **joules per gram** (J/g).

  energy per gram = $\dfrac{\text{energy supplied}}{\text{mass of fuel burnt}}$

  If the water in the previous example has been heated by 3.0 g of fuel, the energy output is:

  energy per gram = $\dfrac{21\,000}{3.0}$ = 7000 J/g

Measuring and calculating energy transferred by fuels.

Questions

Grades D–C
1. What's an exothermic reaction?

Grades B–A*
2. Use ideas about bond breaking and making to explain an endothermic reaction.

Grades D–C
3. What colour flame is seen during incomplete combustion?

Grades B–A*
4. Calculate the energy per gram released by 2.0 g of fuel that raises the temperature of 100 g of water from 20 °C to 50 °C.

C1 Summary

C1 CARBON CHEMISTRY

Food and additives

- All foods are chemicals. Cooking is a **chemical change**. A chemical change is **irreversible** and a **new substance is made**.

- Egg and meat are **proteins**. Proteins in egg and meat change shape when cooked. This process is called **denaturing**.

- **Intelligent packaging** helps with storage of food.

- Food is cooked to:
  – kill microbes
  – improve the **texture**
  – improve **flavour**
  – make it **easier to digest**.

- Potatoes are **carbohydrates**. When cooked, the cell wall breaks so they are easier to digest.

- Additives may be added as:
  – **antioxidants**
  – **colours**
  – **emulsifiers**
  – **flavour enhancers**.

Smells and crude oil

- Perfumes are **esters** that can be made from acids and alcohols.

- Nail varnish doesn't dissolve in water. The attraction between particles in nail varnish is stronger than the attraction between water molecules and molecules in nail varnish.

- The larger the molecules in a fraction, the higher the boiling point. The larger the molecule, the stronger the intermolecular forces between the molecules.

- Crude oil is a fossil fuel made by dead animals being compressed over millions of years. It is **non-renewable**.

- Crude oil is separated by **fractional distillation**. The fractions with the lower boiling points exit at the top of the tower.

- There's not enough petrol made to meet the demand. There's more heavy oil distilled than needed. These larger alkane molecules can be cracked to make smaller, more useful ones, like those of petrol.

Polymers

- **Polymers** are large, long-chain molecules made from small monomers under high pressure with a catalyst.

- **Monomers** are **alkenes**, such as ethene and propene. Alkenes are hydrocarbons made of carbon and hydrogen only.

- Poly(ethene) is used for plastic bags because it is waterproof and flexible. Poly(styrene) is used for packaging and insulation.

- Nylon and Gore-Tex® can be used in coats because they are waterproof. Gore-Tex® has the advantage that it's also breathable. The material is laminated with nylon to make it stronger.

Fossil fuels and energy

- If there's a good supply of oxygen, the products of **complete combustion** of a **hydrocarbon fuel** are **carbon dioxide** and **water**.

- When choosing a fuel to use for a particular purpose, several factors need to be considered:
  – energy value and availability
  – storage and cost
  – toxicity and how much pollution they cause
  – how easy they are to use.

- An **exothermic** reaction **transfers heat out** to the surroundings. An **endothermic** reaction **transfers heat in**.

- Fuels can be compared by heating a fixed amount of water in a calorimeter and measuring the change in temperature. The energy can be calculated by the formula: energy transferred = mass of water × 4.2 × temperature change

Paints and pigments

Paints and pigments

- A paint is a **colloid** where small solid particles are dispersed through the whole liquid, but aren't dissolved.

- When an **oil paint** is painted onto a surface, the solvent **evaporates** leaving the binding medium to dry and form a skin, which sticks the pigment to the surface.

- An **emulsion paint** is a water-based paint. It's made of tiny droplets of one liquid in water, which is called an **emulsion**. When emulsion paint has been painted on to a surface as a thin layer, the water evaporates leaving the binding medium and pigment behind. As it dries it joins together to make a continuous film.

- Oil paint and emulsion paints are colloids because they are a mixture of solid particles in a liquid. The particles don't separate because they're scattered throughout the mixture and are small enough not to settle at the bottom.

- The oil in oil paint is very sticky and takes a long time to harden. Once the solvent has evaporated the oil slowly reacts with oxygen in the air to form a tough, flexible film over the wood. The oil binding medium is **oxidised** by the air.

Thermochromic pigments

- **Thermochromic pigments** are used in paints that are chosen for their colour and also for the temperature at which their colour changes. For example, a thermochromic pigment that changes colour at 45 °C can be used to paint cups or kettles to act as a warning.

- Most thermochromic pigments change from a colour to colourless.

- Thermochromic paints come in a limited range of colours. To get a larger range of colours they are mixed with different colours of normal acrylic paints in the same way that you mix any coloured paints.

- When a green mixture gets hot, the blue thermochromic paint becomes colourless, so all that is seen is the yellow of the acrylic paint.

Phosphorescent pigments

- **Phosphorescent pigments** absorb energy and store it. They can then release it slowly as light. They are sometimes used in luminous clock dials.

- Phosphorescent paints are much safer than radioactive paints that were developed to glow in the dark.

Questions

Grades D-C
1 What is a colloid?

Grades B-A*
2 Describe how oil paint dries and hardens.

Grades B-A*
3 A company is concerned with the type of fumes given off during the drying process when painting. Which type of paint would you recommend, emulsion or oil-based. Explain your answer.

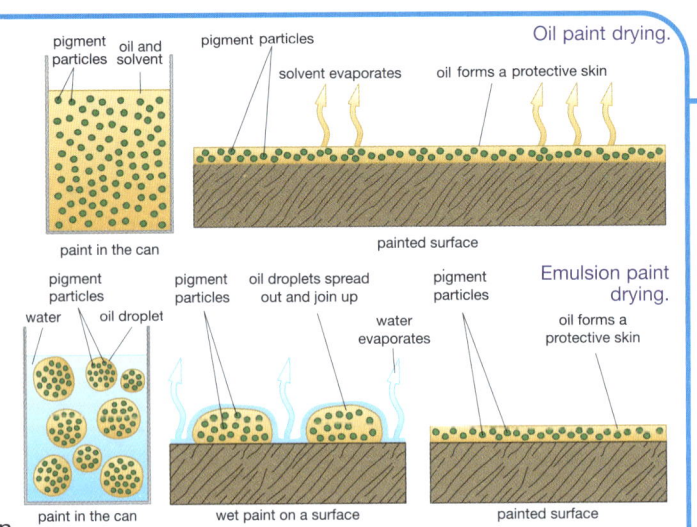

Construction materials

The raw materials

- Limestone is easier to cut into blocks than marble or granite. Marble is much harder than limestone. Granite is harder still and is very difficult to shape.

- Brick, concrete, steel, aluminium and glass come from materials in the ground, but they need to be manufactured from raw materials. This is shown in the table.

raw material	clay	limestone and clay	sand	iron ore	aluminium ore
building material	brick	cement	glass	iron	aluminium

- **Igneous** and **metamorphic** rocks are normally harder than **sedimentary** rocks.
  - Granite is an **igneous rock** and is very hard. Igneous rock is formed out of liquid rock that cools slowly and forms interlocking crystals as it **solidifies**. It is this interlocking structure that gives the rock its hardness.
  - Marble is a **metamorphic rock**. It is not as hard as granite but is harder than limestone. Metamorphic rocks form when a rock has been under **heat and pressure** in the Earth's crust, making it harder than the original rock.
  - Limestone is a **sedimentary rock** and is the softest. Sedimentary rock is made from the shells of dead sea-creatures that stuck together.

Cement and concrete

- **Thermal decomposition** is the chemical breakdown of a compound into at least two other compounds under the effect of heat.

- Calcium carbonate (limestone) thermally decomposes at a very high temperature. This is shown in the word equation:

  calcium carbonate ⟶ calcium oxide + carbon dioxide

- **Cement** is made when limestone and clay are heated together.

- Reinforced concrete has steel rods or steel meshes running through it and is stronger than concrete. It's a **composite** material.

- The symbol equation for the thermal decomposition of calcium carbonate is:

  $CaCO_3 \longrightarrow CaO + CO_2$

- Reinforced concrete is a better construction material than non-reinforced concrete.
  - If a heavy load is put on a concrete beam it will bend very slightly. When a beam bends its underside starts to stretch, which puts it under tension and cracks start to form.
  - Steel is strong under tension. Steel rods in reinforced concrete stop it stretching.

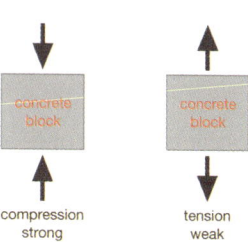

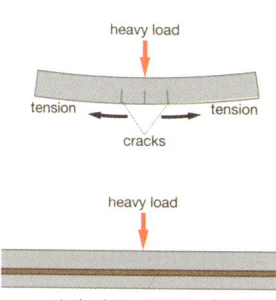

Grades D–C
1 What is cement made from?

Grades B–A*
2 How is limestone changed into marble?

Grades D–C
3 What are the products of heating limestone?

Grades B–A*
4 Write the symbol equation for the thermal decomposition of calcium carbonate.

Does the Earth move?

The structure of the Earth

- The outer layer of the Earth is called the **lithosphere**. It's relatively cold and rigid and is made of the crust and the part of the mantle that lies just underneath.
- The **tectonic plates** that make up the Earth's crust are **less dense** than the mantle and they 'float' on it. There are two kinds of plate:
  - **continental plates** that carry the continents
  - **oceanic plates** that lie underneath the oceans.

  The parts of the lithosphere.

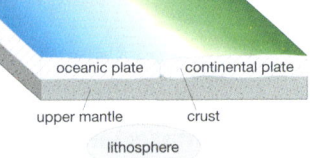

- The crust is far too thick to drill through, so most of our knowledge comes from measuring seismic waves produced by earthquakes. This technology improved in the 1960s when scientists were developing ways of detecting nuclear explosions.

- **Tectonic plates** move very slowly and in different ways: **apart**, **collide**, or **scrape sideways** past each other.
- The **mantle** is hard and rigid near to the Earth's surface and hotter and non-rigid near to the Earth's core. The mantle is always solid, but at greater depths it's more like Plasticine, which can 'flow'.
- In plate tectonics, energy from the hot core is transferred to the surface by slow **convection currents** in the mantle. Oceanic plates are denser than continental plates.
  - When two plates collide, the more dense oceanic plate sinks below the less dense continental plate. This is called **subduction**. The oceanic plate partially re-melts and is reabsorbed into the mantle.
- There have been developments in the theory of plate tectonics.
  - People noticed that the coastline of Africa matches that of South America, suggesting that the continents were formed from one 'supercontinent' which was splitting apart at the time of the dinosaurs (**continental drift theory**).

Top Tip!
Subduction zones are where plates are being destroyed.

Magma, rocks and volcanoes

- Magma rises through the Earth's crust because it's less dense than the crust. It cools and solidifies into **igneous** rock either after it comes out of a volcano as lava, or before it even gets to the surface.
- By looking at crystals of **igneous rock**, geologists can tell how quickly the rock cooled.
  - Igneous rock that **cools rapidly** (close to the surface) has small crystals.
  - Igneous rock that **cools slowly** (further from the surface and better insulated) has large crystals.

- Different types of magma can cause different types of eruption at the surface of the Earth.
  - **Iron-rich magma** (called basaltic magma) is runny and fairly 'safe'. The lava spills over the edges of a volcano and people can escape.
  - **Silica-rich magma** is less runny and produces volcanoes that may erupt explosively. It shoots out as clouds of searingly hot **ash** and **pumice**. The falling ash often includes large lumps of rock called **volcanic bombs**.
- Geologists investigate past eruptions by looking at the ash layers. In each eruption, coarse ash falls first, then fine ash, producing **graded bedding**. Future eruptions can sometimes be predicted with the help of seismometer, but it isn't precise and disasters still occur.

Questions

Grades D-C
1 Why do tectonic plates 'float' on the molten rock?

Grades B-A*
2 Describe the process of subduction.

Grades D-C
3 Explain why the size of crystals changes when magma cools.

Grades B-A*
4 Explain why silica-rich magma is more dangerous than iron-rich magma.

Metals and alloys

Electrolysis

Grades D–C

- Impure copper can be purified in the laboratory using an **electrolysis cell**.
  - The **anode** is impure copper and dissolves into the **electrolyte**.
  - The **cathode** is 'plated' with new copper.

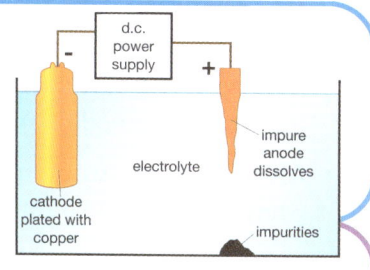

An electrolysis cell.

Grades B–A*

- **Electrolysis** is the break-up of a chemical compound (the electrolyte) when you use an electric current.

- In the purification of copper, the electrolyte is a solution of copper(II) sulfate which is electrolysed using copper electrodes.
  - Instead of the electrolyte breaking apart, the **anode dissolves** and the **cathode** is plated in **pure copper**.
  - A sheet of pure copper is used as a cathode. This sheet gets thicker as more pure copper is plated onto it.
  - The impure copper anode is called **blister copper**. Sometimes it's also called boulder copper.
  - The impurities from the copper anode sink to the bottom of the cell.

Recycling copper

Grades D–C

- Copper has a fairly **low melting point** that makes it easy to melt down and recycle. However, copper that's been already used may be contaminated with other elements, such as solder. This means that it can't be used for purposes where the copper must be very pure, such as electric wiring.

- Copper used for recycling has to be sorted carefully so that valuable 'pure' copper scrap isn't mixed with less pure scrap.

- When impure copper is used to make **alloys**, it must first be analysed to find out how much of each element is present. Very impure scrap copper has to be electrolysed again before it can be used.

Alloys

Grades D–C

- Most **metals** form **alloys**.
  - Amalgam contains mercury.
  - Solder contains lead and tin.
  - Brass contains copper and zinc.

Top Tip! Alloys are often more useful than the metals they're made from.

Grades B–A*

- **Smart alloys** can:
  - be bent more than steel so they're much harder to damage
  - change shape at different temperatures, called '**shape memory**'.

- New ways of using smart alloys are being discovered. Here are some uses.
  - In the frames of glasses to stop them breaking.
  - In shower heads to reduce the water supply if the temperature gets so hot that it scalds.
  - Surgeons can put a small piece of metal into a person's blocked artery and then warm it slightly. As it warms up, it changes shape into a much larger tube that holds the artery open and reduces the risk of a heart attack.

- **Nitinol** is a smart alloy made from nickel and titanium.

Questions

Grades D-C
1. Which electrode becomes pure copper in electrolysis?

Grades B-A*
2. In the purification of copper, what's the electrolyte?

Grades D-C
3. What metals form the alloy brass?

Grades B-A*
4. Which property do smart alloys have that depends on temperature?

C2 ROCKS AND METALS

Cars for scrap

Rusting and corrosion

- **Acid rain** and **salt water** accelerate **rusting**. In winter, icy roads are treated with salt, which means that car bodies rust quicker.
- **Aluminium** doesn't corrode in moist air because it has a protective layer of **aluminium oxide** which doesn't flake off the surface.

- Rusting is a **chemical reaction** between iron, oxygen and water called **oxidation**. This is because iron reacts with oxygen to make an oxide.
- The chemical name for rust is **hydrated iron(III) oxide**.
- The word equation for rusting is:

  iron + water + oxygen ⟶ hydrated iron(III) oxide

Materials used in cars

- **Alloys** often have different and more useful properties than the pure metals they're made from. **Steel** (an alloy made of **iron** and **carbon**) is stronger and harder than iron and doesn't rust as easily as pure iron.
- Steel and aluminium can both be used to make **car bodies**, but each material has its advantages.

steel	aluminum
stronger and harder than aluminium which is important in the event of a crash	mass of a car body is less than the same car body made from steel, so has a better **fuel economy**
car body is cheaper than one made of aluminium	aluminium car body will corrode less so the car body has a much **longer lifetime**

Top Tip! For grades D-C and B-A*, you need to know the information in this table.

Recycling

- More recycling of **metals** means that less metal ore needs to be mined.
- Recycling of **iron** and **aluminium** saves money and energy compared to making them from their ores.
- Recycling **plastics** means less crude oil is used and less non-biodegradeable waste is dumped.
- Recycling of **glass** has been happening for many years very successfully.
- Recycling **batteries** reduces the dumping of toxic materials into the environment.
- European Union law requires 85% of a car to be recyclable. This percentage will increase to 95% in the future. Technology has to be developed to separate all the different materials used in making a car.

Questions

Grades D-C
1. Explain how aluminium is protected from corrosion in moist air.

Grades B-A*
2. What's the chemical name for rust?

Grades D-C
3. Why is the fuel economy better in an aluminium car rather than one of steel?

Grades B-A*
4. Give one reason why a car should be made from aluminium and one why it should be made of steel.

Clean air

What's in clean air?

- Clean air is made up of 78% nitrogen, 21% oxygen and of the remaining 1%, only 0.035% is carbon dioxide.

- These percentages don't change very much because there's a balance between processes that use up and make carbon dioxide and oxygen. These processes are shown in the **carbon cycle**.

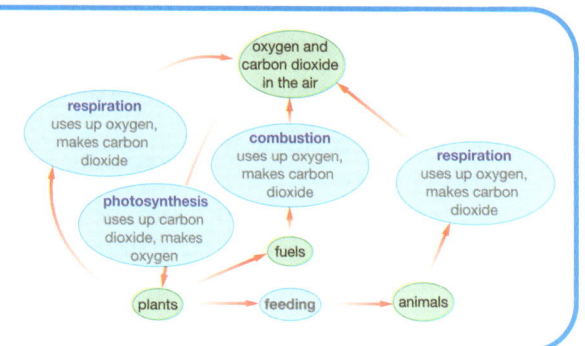

- Over the past few centuries the percentage of carbon dioxide in the air has increased due to:
  – **increased energy usage** – more fossil fuels are being burnt in power stations
  – **increased population** – the world's energy requirements increase
  – **deforestation** – as more rainforests are cut down, less photosynthesis takes place.

The atmosphere

- Scientists **know** that gases trapped in liquid rock under the surface of the Earth are always escaping. This happens in volcanoes.

- Scientists **guess** about the original atmosphere of the Earth. It's known that microbes developed that could photosynthesise. These organisms could remove carbon dioxide from the atmosphere and add oxygen. Eventually the level of oxygen reached what it is today.

- Scientists **know** that the gases came from the centre of the Earth in a process called **degassing**.

- They **think** that the original atmosphere contained ammonia and later carbon dioxide. A chemical reaction between ammonia and rocks produced nitrogen and water. The percentage of nitrogen slowly increased, and as it is unreactive, very little nitrogen was removed. Much later, organisms that could photosynthesise evolved and converted carbon dioxide and water into oxygen. As the percentage of oxygen in the atmosphere increased, the percentage of carbon dioxide decreased, until today's levels were reached.

Pollution control

- You need to be able to describe the origin of these atmospheric pollutants.

pollutant	carbon monoxide	oxides of nitrogen	sulfur dioxide
origin of pollutant	incomplete combustion of petrol or diesel	formed in the internal combustion engine	sulfur impurities in fossil fuels

- A car fitted with a **catalytic converter** changes carbon monoxide into carbon dioxide.

- A reaction between nitric oxide and carbon monoxide takes place on the surface of the catalyst. The two gases formed are natural components of air.

  carbon monoxide + nitric oxide ⟶ nitrogen + carbon dioxide
  $2CO + 2NO \longrightarrow N_2 + 2CO_2$

Questions

Grades D–C
1. Which process in the carbon cycle uses up carbon dioxide?

Grades B–A*
2. Explain why deforestation causes problems for the atmosphere.

Grades D–C
3. What does a catalytic converter do?

Grades B–A*
4. Write down the equation for removing carbon monoxide from the atmosphere using a catalytic converter.

Faster or slower (1)

Simple collision theory

- The more collisions there are in a reaction, the faster the reaction.
- The rate of a chemical reaction can be increased by increasing the **concentration** and **temperature**.
  - As the concentration increases the particles become more crowded. This means there are more collisions so the rate of reaction speeds up.
  - The reacting particles have more **kinetic energy** and so the number of collisions increases.
- **Collision frequency**, *not* the number of collisions, determines the rate of a reaction. This describes the number of **successful collisions** between reactant particles each second.
  - For a successful collision, each particle must have lots of kinetic energy.
  - As the concentration increases, the number of collisions per second increases and so the rate of reaction increases.
  - As the temperature increases, the reactant particles have more kinetic energy so there are more energetic collisions. These more energetic collisions are more successful.

Rates of reaction

- Magnesium ribbons of hydrochloride acid were reacted in some experiments.

Graph A shows how rate of reaction changes with a change in **concentration** of the reactants.

Graph B shows how rate of reaction changes with a change in **temperature**.

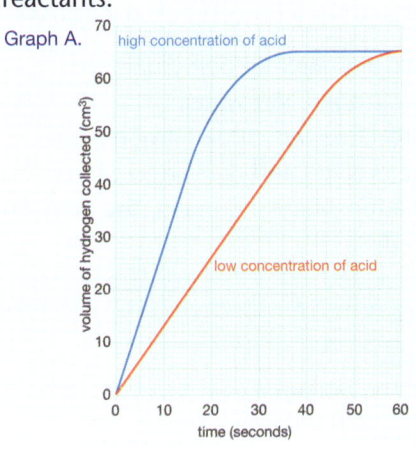

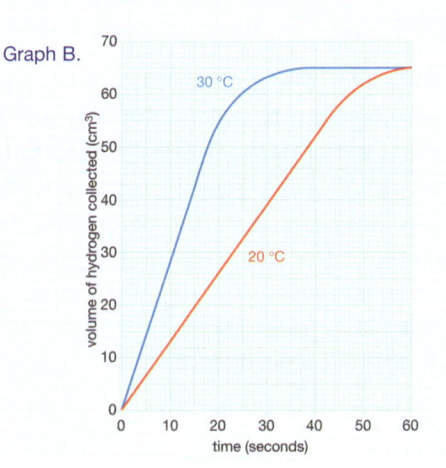

- The total volume of hydrogen produced during all experiments is the same because excess acid and the same mass of magnesium are used.

- The rate of reaction can be worked out from the **gradient** of a graph, which can be found by drawing construction lines.
  - Choose a part of the graph where there's a straight line (not a curve).
  - Measure the value of y and x.
  - Then divide y by x.
  - The gradient of the graph is: gradient = $\frac{y}{x}$

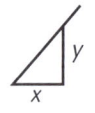

Top Tip!
Extrapolation means extending the graph to read it. **Interpolation** means reading within the graph between closer points.

Questions

Grades D-C
1 Use ideas about particles to explain how the rate of a reaction can be altered.

Grades B-A*
2 Explain what's meant by 'collision frequency'.

Grades D-C
3 If a reaction between the same mass of magnesium and excess acid is measured at two different temperatures, the total volume of gas produced doesn't change. Explain why.

Grades B-A*
4 What does 'extrapolation' mean?

Faster or slower (2)

Explosions

- During an **explosion**, a large volume of gaseous products are released, moving outwards from the reaction at great speed causing the explosive effect.
- **Combustible powders** often cause explosions.
  - A powder reacts with oxygen to make large volumes of carbon dioxide and water vapour.
  - A factory using combustible powders such as sulfur, flour, custard powder or even wood dust must be very careful. The factory owners must ensure that the powders can't reach the open atmosphere and that the chance of producing a spark is very small.

Surface area

- A powdered reactant has a much larger **surface area** than the same mass of a block of reactant.
- As the surface area of a solid reactant increases so does the rate of reaction.
- Fewer reacting particles of **B** can be in contact with reacting particles of **A**. As the surface area increases there are more collisions, which means the rate of reaction increases.
- The graph shows the rate of reaction between calcium carbonate and dilute hydrochloric acid.
  - As the reaction takes place, the mass on the balance decreases. This is because carbon dioxide gas is escaping.
  $CaCO_3 + 2HCl \longrightarrow CaCl_2 + H_2O + CO_2$
  - The **gradient** of the graph is a measure of the rate of reaction. As the reaction takes place, the rate of reaction becomes less and less because the concentration of acid and the mass of calcium carbonate decrease.
  - As the reaction proceeds there are fewer collisions between reactants.
- It's the **collision frequency** between reactant particles that's important in determining how fast a reaction takes place. The more successful collisions there are each second, the faster the reaction.
- When the surface area of a solid reactant is increased, there will be more collisions each second. This means the rate of reaction increases.

Catalysts

- A **catalyst**:
  - increases the rate of a reaction
  - is unchanged at the end of a reaction
  - is needed in small quantities to catalyse a large mass of reactants
  - usually only makes a **specific** reaction faster
  - doesn't increase the number of collisions per second
  - works by making the collisions more successful
  - helps reacting particles collide with the correct orientation
  - allows collisions between particles with less kinetic energy than normal to be successful.

Questions

Grades D–C
1. Give two examples of combustible powders.
2. Why is only 50 cm³ of hydrogen gas produced when using a 0.135 g lump of zinc with acid, compared to 100 cm³ of gas produced when using a 0.27 g lump of zinc?

Grades B–A*
3. What happens to the number of particle collisions if the surface area of a reactant is increased?
4. How does a catalyst help speed up a reaction?

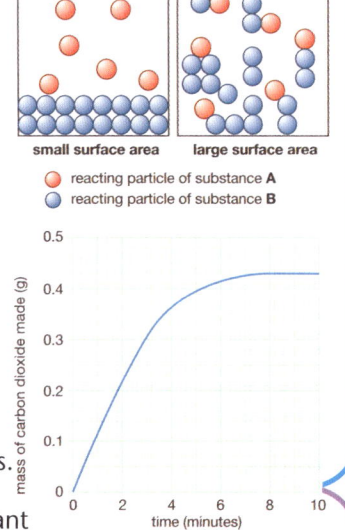

C2 Summary

C2 ROCKS AND METALS

Paints and pigments

- Paints are made from:
  - coloured rock (**pigment**) particles
  - **solvent**
  - **binding medium**.

- Paint is a **colloid**. Oil paints 'dry' by oxidation with atmospheric oxygen.

- Paints are applied as thin layers, which dry when the solvent evaporates.

Building materials

- These are manufactured from rocks:
  - brick is made from clay
  - glass is made from sand
  - aluminium or iron are made from their ores.

- When limestone is heated to thermally decompose it, one substance is changed into two: calcium oxide and carbon dioxide.

Metals

- Metals from cars are easy to recycle. Plastics need legislation.

- Pure metals can be mixed with other elements to make **alloys**:
  - copper and zinc make brass
  - lead and tin make solder.

- Aluminium doesn't corrode in moist conditions because it has a protective layer of aluminium oxide which doesn't flake on the surface.

- Iron and aluminium are used in cars. Iron rusts, but aluminium doesn't. **Rusting** is the oxidation of iron to form hydrated iron oxide.

- Alloys have properties that make them more useful than the pure metal. For example, steel is harder and stronger than iron.

Earth and atmosphere

- The Earth is made of **tectonic plates** that float on the mantle. The plates are moving all the time.

- Tectonic plates are found on top of the mantle as they're less dense than the mantle.

- If molten rock cools slowly the crystals that are formed are bigger.

- The atmosphere used to be poisonous, but then plants produced oxygen. The composition of the atmosphere is now 21% oxygen and 78% nitrogen.

- Carbon dioxide is given out in **combustion** but taken in during **photosynthesis**.

- The Earth's original atmosphere came from gases escaping from the interior of the Earth.

Fast or slow?

- Rates of reaction are affected by:
  - **temperature**
  - **surface area**
  - **concentration**
  - **catalysts**.

- An increase in surface area of a reactant increases the frequency of collisions.

- The higher the temperature, the faster the particles move. This increases the rate of reaction.

What are atoms like?

Atoms

- The nucleus of an atom is made up of **protons** and **neutrons**.
- The **atomic number** is the number of protons in an atom.
- The **mass number** is the total number of protons and neutrons in an atom.

	relative charge	relative mass
electron	−1	0.0005 (zero)
proton	+1	1
neutron	0	1

Helium has two protons (red) and two neutrons (green). Helium has a mass number of 4.

A proton has a positive charge. An electron has a negative charge. This atom is neutral because it has the same number of protons and electrons. (Neutrons not shown.)

- An atom is **neutral** because it has an equal number of electrons and protons. The positive charges balance out the negative charges.
- If a particle has an atomic number of 11, a mass number of 23 and a neutral charge, it must be a sodium atom.

atomic number	11
mass number	23
charge	0
	sodium atom

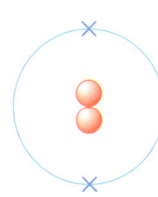

- An element has the symbol $^{14}_{6}C$, it has no charge so it's an atom. It has 6 protons, a mass number of 14 and must therefore have 8 (14 − 6) neutrons. $^{14}_{6}C$ is sometimes written as carbon-14.

Top Tip!
The elements in the periodic table are arranged in **ascending** atomic number.

Isotopes

- Isotopes are elements that have the same atomic number but different mass numbers.
- Isotopes of an element have different numbers of neutrons in their atoms.

isotope	electrons	protons	neutrons
$^{1}_{1}H$	1	1	0
$^{2}_{1}H$	1	1	1
$^{3}_{1}H$	1	1	2

Electronic structure

- The same number of electrons occupies the space around the protons of the nucleus.
- Electrons occupy **shells**. The electron shell nearest to the nucleus takes up to 2 electrons. The second shell takes up to 8 electrons.
- Each element has an **electron pattern** (electronic structure).
  - The **third shell** takes up to 8 electrons before the fourth shell starts to fill.
  - The **fourth shell** can take up to 18 electrons.
- The electronic structure of each of the first twenty elements can be worked out using:
  - the atomic number of the element
  - the maximum number of electrons in each shell.

The distribution of electrons in **a** a helium atom **b** a carbon atom.

the first shell takes up to 2 electrons

the second shell takes up to 8 electrons

$^{4}_{2}He$

$^{12}_{6}C$

The atomic number of lithium, Li, is 3. So the first 2 electrons of the lithium atom fill the first shell. The third electron goes into the second shell. The electronic structure is 2,1.

Questions

Grades D-C
1 What's the mass of one proton?

Grades B-A*
2 What's the isotope that has 17 protons and 20 neutrons?

Grades D-C
3 How many electrons does an atom of sodium have?

Grades B-A*
4 What's the element with electronic structure 2,8,2?

Ionic bonding

Ionic bonding

- A **metal atom** has extra electrons in its outer shell and needs to **lose** them to be stable. The electrons transfer from the metal atom to a non-metal atom to form a stable pair.
  - If an atom **loses electrons**, a **positive ion** is formed.
  - If an atom loses 1 electron, a (positive) $^+$ion is formed, e.g. Na – e$^-$ ⟶ Na$^+$
  - If an atom loses 2 electrons, a (positive) 2$^+$ ion is formed, e.g. Mg – 2e$^-$ ⟶ Mg^{2+}

- A **non-metal** atom has 'spaces' in its outer shell and needs to **gain** electrons to be stable. The electrons transfer to the non-metal atom from the metal atom to make a stable pair.
  - If an atom **gains electrons**, a **negative ion** is formed.
  - If an atom gains 1 electron, a (negative) $^-$ion is formed, e.g. F + e$^-$ ⟶ F$^-$
  - If an atom gains 2 electrons, a (negative) 2$^-$ ion is formed, e.g. O + 2e$^-$ ⟶ O^{2-}

- During **ionic bonding**, the metal atom becomes a positive ion and the non-metal atom becomes a negative ion. The positive ion and the negative ion then attract one another.

- Sodium chloride **solution** conducts electricity. **Molten** (melted) magnesium oxide and sodium chloride conduct electricity.

- The '**dot and cross**' model describes ionic bonding. **Sodium chloride** is a solid lattice made up of many pairs of ions held together by **electrostatic** attraction.
  - Sodium forms a positive sodium ion.
  - Chlorine forms a negative chloride ion.
  - The outer electron of sodium is transferred to the outer shell of the chlorine atom. 
  - The sodium ion and the chloride ion are held together by attraction of opposite charges.

- In **magnesium chloride**, magnesium needs to lose two electrons but chlorine can only gain one electron. So one magnesium atom needs two chlorine atoms to achieve an ionic bond and make magnesium chloride.

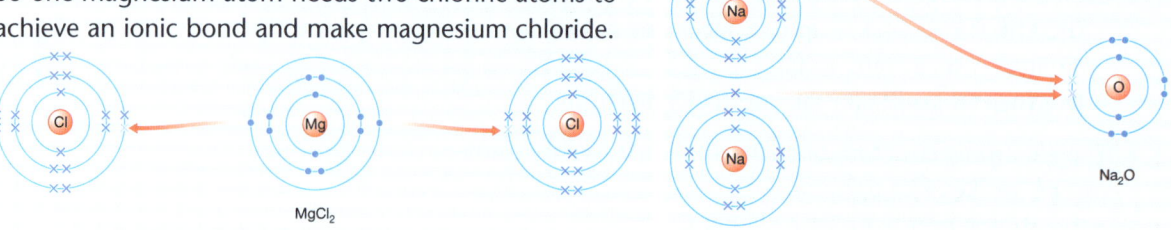

- During the bonding in **magnesium oxide**, the magnesium atom loses two electrons to become a magnesium ion and the oxygen gains the two electrons to become an oxide ion. 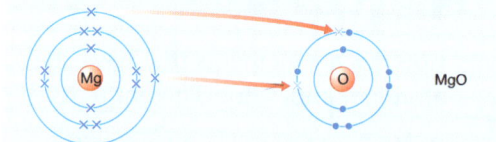
  - The dot and cross model looks like this:

- In **sodium oxide**, sodium only has one electron to lose, but oxygen needs to gain two electrons. Two sodium atoms are needed to bond with one oxygen atom.

- Each atom has either lost or gained the correct number of electrons to achieve a complete outer shell. It's called a **stable octet**.

- Molten liquids of ionic compounds can conduct electricity as the ions are free to move.

Questions

Grades D-C

1. Explain how a negative ion is made from a neutral atom.
2. Explain the difference between how a metal atom and a non-metal atom transfers electrons.

Grades B-A*

3. Draw a 'dot and cross' model to show how an electron is transferred from a lithium atom to a fluorine atom.
4. Why do molten liquids of ionic compounds conduct electricity?

Covalent bonding

Covalent bonding

Grades D–C

- Non-metals combine together by sharing **electrons**. This is called **covalent bonding**.

  A molecule of water is made up of three atoms: two hydrogen and one oxygen.
  - Oxygen has six electrons in its outer shell; it needs two more electrons to be complete.
  - Hydrogen atoms each have one electron in their only shell, so the oxygen outer shell is shared with each of the hydrogen electrons.
  - So each of the hydrogen atoms has a share of two more electrons making the shell full. 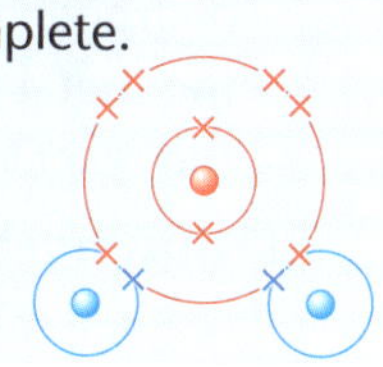

  A molecule of carbon dioxide is made up of three atoms: two oxygen and one carbon.
  - Carbon has four electrons in its outer shell; it needs four more electrons to be complete.
  - Oxygen atoms each have six electrons in their outer shell, so they each need two more electrons to be complete.
  - The oxygen outer shell is shared with two of the electrons of the carbon outer shell each.
  - So each of the oxygen atoms has a share of two more electrons making the shell full.

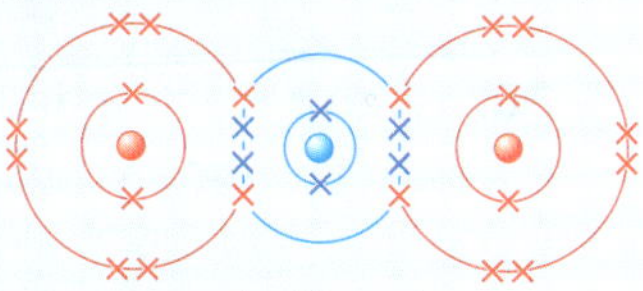

- Carbon dioxide and water don't conduct electricity because they are covalently bonded.

Grades B–A*

- The formation of simple molecules containing single and double covalent bonds can be represented by **'dot and cross' models**.

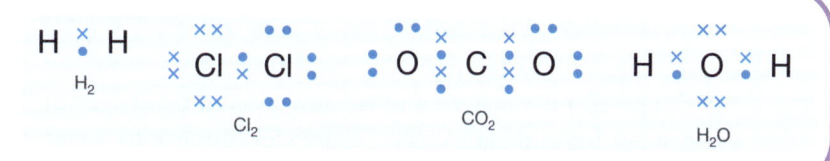

Predicting chemical properties

Grades B–A*

- Carbon dioxide and water are simple molecules with weak **intermolecular forces**.

- The chemical properties of carbon dioxide and water are related to their structure.
  - As they have weak intermolecular forces between the molecules, they're easy to separate, so the substances have low melting points.
  - As there are no free electrons available they don't conduct electricity.

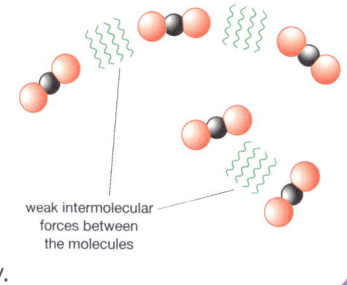

Group numbers

Grades D–C

- This is how to tell which **group number** an element belongs to:
  - group 1 elements have 1 electron in the outer shell
  - group 7 elements have 7 electrons in the outer shell
  - group 8 elements have 8 electrons in the outer shell.

- This is how to tell which **period** an element belongs to:
  - electrons in only **one shell**, it's in the **first period**
  - electrons in **two shells**, it's in the **second period**
  - electrons in **three shells**, it's in the **third period**.

element	electrons	period
H	1	1
Li	2,1	2
Na	2, 8, 1	3

Questions

Grades D–C

1 Draw a model showing the bonding of a water molecule.

Grades B–A*

2 Draw a dot and cross model of the bonding in water.

Grades D–C

3 Sulfur has an electron pattern of 2, 8, 6. To which period does it belong?

Grades B–A*

4 There are only weak intermolecular forces between the molecules of carbon dioxide. Explain why this causes it to have a low melting point.

The group 1 elements

Properties of alkali metals

- When lithium, sodium and potassium react with water:
  - they float on the surface because their **density** is less than the density of water
  - hydrogen gas is given off
  - the metal reacts with water to form an **alkali** – the **hydroxide** of the metal
  - the reactivity of the alkali metals with water increases down group 1.

  Lithium reacts quickly and vigorously with water.

  lithium + water ⟶ lithium hydroxide + hydrogen

  Sodium reacts very quickly and vigorously with water and forms sodium hydroxide.

  sodium + water ⟶ sodium hydroxide + hydrogen

  Potassium reacts extremely vigorously with water and produces a lilac flame and forms potassium hydroxide.

  potassium + water ⟶ potassium hydroxide + hydrogen

- Group 1 metals have **1 electron** in their **outer shell**, which is why they have similar properties.

- The word equation: sodium hydroxide + water ⟶ sodium + hydrogen can be represented in symbols: $Na + H_2O \longrightarrow NaOH + H_2$ but the equation isn't balanced.
  - The balanced symbol equation is: $2Na + 2H_2O \longrightarrow 2NaOH + H_2$

- The reactivity with water increases down group 1.
  - We can predict the way that other alkali metals behave from the patterns so far.
  - We can see that rubidium reacts more vigorously with water than potassium.
  - We can predict the trends of the melting point, boiling point, appearance and electrical conductivity of rubidium and caesium from the trends of the other three.

$_3Li$
$_{11}Na$
$_{19}K$
$_{37}Rb$
$_{55}Cs$
$_{87}Fr$

Flame tests

- If you want to test the flame colours of the chemicals:
  - Put on safety goggles. Moisten a flame test wire with dilute hydrochloric acid.
  - Dip the flame test wire into the sample of solid chemical.
  - Hold the flame test wire in a blue Bunsen burner flame.
  - Record the colour of the flame in a table.

Explaining reactivity patterns

- Alkali metals have similar properties because when they react, their atoms need to lose one electron to form full outer shells. This is then a **stable electronic structure**.
  - When the atom loses one electron it forms an **ion**. The atom becomes charged.
  - It has more positive charges in its nucleus than negative electrons surrounding it, so the overall charge is positive. It has made a **positive ion**.
  - This can be represented by an equation. $Na - e^- \longrightarrow Na^+$

- Lithium loses its outer electron from its second shell. Sodium loses its outer electron from its third shell. The third shell is further away from the attractive 'pulling force' of the nucleus so the electron from sodium is more easily lost than the electron from lithium. Sodium is therefore more reactive than lithium.

- If electrons are lost, the process is called **oxidation**.

Questions

Grades D–C

1. Why does potassium float on water?
2. Explain why group 1 metals have similar properties.

Grades B–A*

3. What alkali will be made when rubidium reacts with water?
4. Why is sodium less reactive than potassium?

The group 7 elements

Group 7 trends

- The **physical appearance** of the halogens at room temperature is:
  - chlorine is a green gas
  - bromine is an orange liquid
  - iodine is a grey solid.

- Group 7 elements have similar properties:
  - they all have **7 electrons** in their **outer shell**.

  - melting points and boiling points increase down the group
  - when they react each atom gains one electron to form a **negative ion** with a stable electronic structure.

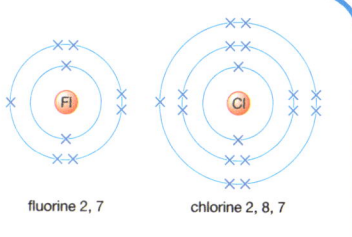

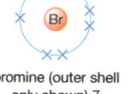

  Chlorine has an electronic structure of 2, 8, 7. It gains one electron to become 2, 8, 8.
  $$Cl_2 + 2e^- \longrightarrow 2Cl^-$$

- The nearer the outer shell is to the nucleus, the easier it is for an atom to gain one electron. The easier it is to gain the electron, the more reactive the halogen.

- If electrons are gained, the process is called **reduction**. In the ionic equation above, a molecule of chlorine gains two electrons to become two chlorine ions. The chlorine ion is a negative ion.

fluorine 2, 7
chlorine 2, 8, 7
bromine (outer shell only shown) 7
iodine (outer shell only shown) 7

Halogens and reactivity

- Halogens react vigorously with alkali metals to make **metal halides**.

  When lithium reacts with chlorine, the metal halide made is lithium chloride.
  lithium + chlorine ⟶ lithium chloride

  When potassium reacts with iodine, the metal halide made is potassium iodide.
  potassium + iodine ⟶ potassium iodide

- To construct a **balanced symbol equation**:
  - write down the symbols for the alkali metal and the halogen (the **reactants**)   potassium K (alkali metal) chlorine, Cl_2 (halogen)
  - write down the formula for the **product** $K + Cl_2 \longrightarrow KCl$
  - balance number of molecules $2K + Cl_2 \longrightarrow 2KCl$

Displacement reactions of halogens

- The **reactivity** of the halogens decreases down the group.

- If halogens are bubbled through **solutions of metal halides**, there are two possibilities:
  - **no reaction**: if the halogen is less reactive than the halide in solution
  - a **displacement reaction**: if the halogen is more reactive than the halide in solution.

  Chlorine displaces the bromide to form bromine solution.
  chlorine + potassium bromide ⟶ potassium chloride + bromine (orange solution)

  Chlorine also displaces iodides from sodium iodide solution.
  chlorine + sodium iodide ⟶ sodium chloride + iodine (red-brown solution)

Questions

Grades B-A*
1. What's reduction?

Grades D-C
2. Write down the word equation for the reaction between potassium and bromine.

Grades B-A*
3. Write a balanced symbol equation for the formation of lithium bromide from its elements.

Grades D-C
4. Why doesn't iodine displace bromine from potassium bromide?

Electrolysis

The electrolysis of dilute sulfuric acid

- The key features of the **electrolysis** of dilute sulfuric acid are:
  - the **electrolyte** is a dilute solution of sulfuric acid
  - two electrodes are connected to a DC supply, between 6 V and 12 V, and placed into the electrolyte
  - the electrode connected to the **negative terminal** is the **cathode**
  - the electrode connected to the **positive terminal** is the **anode**.

- When the current is switched on, bubbles of gas appear at both electrodes. Water splits into two ions: H^+ is the positive ion and OH^- is the negative ion.
  - H^+ is attracted to the negative cathode and discharged as hydrogen gas, H_2.
  - OH^- is attracted to the positive anode and discharged as oxygen gas, O_2.

- Twice the volume of hydrogen gas is given off as oxygen gas because the formula of the compound breaking up is H_2O.

Electrolysis of sulfuric acid in the laboratory.

Electrolysis of sodium chloride

- The reactions at the electrodes in the **electrolysis** of a dilute solution of **sodium chloride** are:
  - **at the cathode**: $2H^+ + 2e^- \longrightarrow H_2$
  - **at the anode**: $4OH^- - 4e^- \longrightarrow 2H_2O + O_2$

Electrolytic decomposition

- The key features in the production of aluminium by electrolytic decomposition are:
  - the use of molten aluminium oxide
  - aluminium is formed at the graphite cathode; oxygen is formed at the graphite anode
  - the anodes are gradually worn away by **oxidation**
  - the process requires a high electrical energy input.

- The word equation for the decomposition of aluminium oxide is:

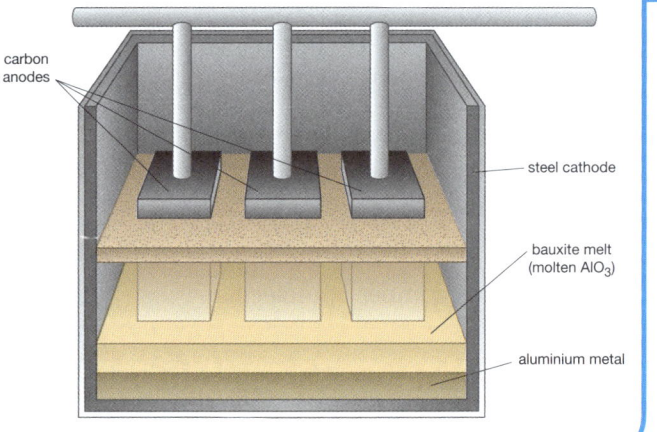

  aluminium oxide $\longrightarrow$ aluminium + oxygen

- The electrode reactions in the **electrolytic** extraction of **aluminium** are:
  - **at the cathode**: electrons have been gained – an example of **reduction**
    $Al^{3+} + 3e^- \longrightarrow 3Al$
  - **at the anode**: electrons have been lost – an example of **oxidation**
    $2O^{2-} - 4e^- \longrightarrow 2O_2$

- The chemical **cryolite** is used to lower the melting point of aluminium oxide. Aluminium oxide requires large amounts of electricity to melt at very high temperatures, which is very expensive.

Questions

Grade D-C

1. The ratio of hydrogen gas to oxygen gas made during the electrolysis of water is 2:1. Explain why.

Grade B-A*

2. The H^+ ions are discharged at the cathode as H_2. Explain how.

Grade D-C

3. What happens to the anodes during the process of electrolysis of aluminium?

Grade B-A*

4. Why is cryolite added in the electrolysis of aluminium oxide?

Transition elements

Transition elements

- A **compound** that contains a transition element is often coloured:
  - copper compounds are blue
  - iron(II) compounds are pale green
  - iron(III) compounds are orange/brown.

- A transition metal and its compounds are often **catalysts**:
  - iron is used in the **Haber process** to make ammonia, which is used in fertilisers
  - nickel is used to harden the oils in the manufacture of margarine.

Top Tip! A catalyst is an element or compound that changes the rate of a chemical reaction without taking part in the reaction. Catalysts are unchanged during the reaction.

Sodium hydroxide solution

- Sodium hydroxide solution is used to identify the presence of transition metal ions in solution:
  - Cu^{2+} ions form a blue solid
  - Fe^{2+} ions form a grey/green solid
  - Fe^{3+} ions form an orange **gelatinous** solid.

- These solids are metal hydroxide precipitates.

What are the precipitates in these test tubes?

Thermal decomposition

- If a transition metal carbonate is heated, it **decomposes** to form a metal oxide and carbon dioxide. On heating:
  - $FeCO_3$ decomposes forming iron oxide and carbon dioxide
  - $CuCO_3$ decomposes forming copper oxide and carbon dioxide
  - $MnCO_3$ decomposes forming manganese oxide and carbon dioxide
  - $ZnCO_3$ decomposes forming zinc oxide and carbon dioxide.

- The metal carbonates change colour during decomposition.

Top Tip! The test for carbon dioxide is that it turns limewater milky.

- To write the balanced symbol equation for thermal decomposition:
  - write a word equation to establish the products of the reaction:

    copper carbonate ⟶ copper oxide + carbon dioxide

  - assign symbols to the words:

    $FeCO_3 \longrightarrow FeO + CO_2$
    $CuCO_3 \longrightarrow CuO + CO_2$
    $MnCO_3 \longrightarrow MnO + CO_2$
    $ZnCO_3 \longrightarrow ZnO + CO_2$
    (these are balanced)

  - sometimes you need to balance the equation:

    $Cu^{2+} + 2OH^- \longrightarrow Cu(OH)_2$
    $Fe^{2+} + 2OH^- \longrightarrow Fe(OH)_2$
    $Fe^{3+} + 3OH^- \longrightarrow Fe(OH)_3$

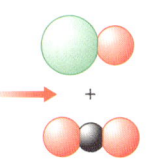

Questions

Grades D-C
1. What colour are iron(III) compounds?
2. There is a difference between Fe^{2+} ions and Fe^{3+} ions. How would you show this using sodium hydroxide?

Grades B-A*
3. Write a word equation for the thermal decomposition of copper carbonate.
4. Write an ionic equation for the precipitation reaction between copper ions and hydroxide ions.

Metal structure and properties

Properties of metals

- A property can be either **physical** or **chemical**.

- Physical properties of metals include:
  – having high thermal conductivity
  – being good conductors of heat
  – being malleable
  – being ductile
  – having high melting points and boiling points because of strong metallic bonds.

  Copper is often used for the base or the whole of saucepans because it has high thermal conductivity.

- Chemical properties of metals include:
  – resistance to attack by oxygen or acids.

  Copper is also resistant, which is another reason why it's used for saucepans.

- Aluminium has a low density and is used where this property is important, such as in the aircraft industry and also in modern cars.

- A **metallic bond** is a strong electrostatic force of attraction between close-packed positive metal ions and a 'sea' of **delocalised electrons**.

- Metals have high melting points and boiling points because a lot of energy is needed to overcome the strong attraction between the delocalised electrons and the positive metal ions.

Conductors and superconductors

- When metals conduct electricity, the electrons in the metal move. **Superconductors** are materials that conduct electricity with little or no resistance.

- When a substance goes from its normal state to a superconducting state, it no longer has a magnetic field. This is called the Meissner effect.

- The potential benefits of superconductors are:
  – loss-free power transmission
  – super-fast electronic circuits
  – powerful electromagnets.

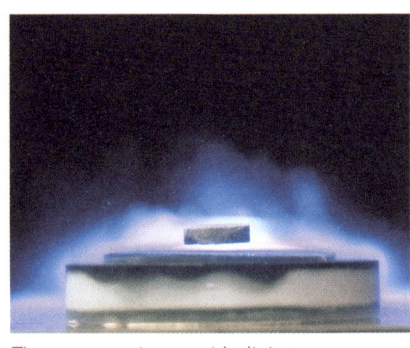

The permanent magnet levitates above the superconductor.

- Metals conduct electricity because delocalised electrons within its structure can move easily.

- Superconductors only work at very low temperatures, so scientists need to develop superconductors that will work at 20 °C.

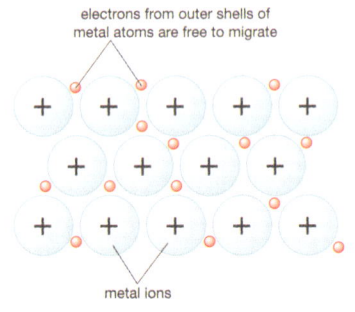

The structure of a metal.

Questions

Grades D-C

1. Give two reasons why copper is good for making saucepans.
2. A metallic bond is a strong bond. Explain why.
3. What happens if a small permanent magnet is put above a superconductor?

Grades B-A*

4. What are the disadvantages of superconductors at present?

C3 Summary

Atoms and bonding

Atoms are neutral because they have the same number of electrons as protons. The electrons are arranged in a pattern or configuration. The outer shell of electrons needs to be full to be stable.

Atoms join together to make molecules or large crystal structures. There are two ways in which atoms can bond, by making ions or by sharing electrons.

Ions are made when atoms lose or gain electrons to make a full outer shell. Ions are either positive or negative.

Atoms can share electrons to make molecules containing two or more atoms. This bonding is called **covalent bonding**.

Periodic table

Group 1 metals react vigorously with water to make alkaline solutions, which are the hydroxide of the metal.

Group 7 elements are called the halogens. They have seven electrons in their outer shell.

The periodic table lists all elements in order of their **atomic number**. The elements of group 1 all have one electron in their outer shell.

The periodic table lists elements in groups. The elements have similar properties. The group that an element belongs to can be deduced from its electron pattern.

Electrolysis

Sulfuric acid decomposes to give hydrogen and oxygen. Hydrogen is made at the cathode and oxygen is made at the anode.

Electrolysis is the decomposition of a substance using electricity.

Aluminium is made by the electrolysis of bauxite. The mineral has to be purified before it's used. The aluminium is deposited at the cathode. Oxygen is given off at the anode. The graphite anodes are burned away in the process.

Transition metals and metal structure

Transition metals and their compounds are often catalysts. Their carbonates decompose on heating to give the **metal oxide** and **carbon dioxide**.

Metals **conduct electricity** easily because electrons move through the structure easily. At low temperatures some metals can become **superconductors**. These show little or no resistance when conducting electricity.

Transition metal compounds are usually coloured. The compounds often dissolve in water to make coloured solutions. The solutions react with sodium hydroxide to make coloured precipitates.

C3 THE PERIODIC TABLE

Acids and bases

Neutralising acids

- An alkali is a **base** which dissolves in water.
- The word equation for **neutralisation** is: acid + base ⟶ salt + water
- Metal oxides and metal hydroxides neutralise acids because they're bases. The reaction of a metal oxide or a metal hydroxide with an acid is:
  acid + oxide ⟶ salt + water   acid + hydroxide ⟶ salt + water
- **Carbonates** also neutralise acids to give water and a gas:
  acid + carbonate ⟶ salt + water + carbon dioxide
- A **salt** is made from part of a base and part of an acid.
- To work out the name of a salt, look at the acid and base it was made from. The first part of the salt name is from the base and the second part from the acid.

Top Tip! Nitrates come from nitric acid, chlorides come from hydrochloric acid, sulfates come from sulfuric acid.

When sodium hydroxide reacts with hydrochloric acid, the salt formed is sodium chloride:

sodium from the base   **chloride** from the acid

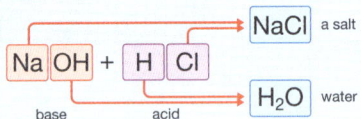

Na OH + H Cl ⟶ NaCl (a salt) + H_2O (water)

- The following are the **balanced symbol equations** for the reactions between some common acids and bases.

  $HCl + NaOH \longrightarrow NaCl + H_2O$ $2HNO_3 + CuO \longrightarrow Cu(NO_3)_2 + H_2O$
  $H_2SO_4 + CaCO_3 \longrightarrow CaSO_4 + H_2O + CO_2$ $2HCl + CaCO_3 \longrightarrow CaCl_2 + H_2O + CO_2$

Hydrogen ions

- An **acid solution** contains **hydrogen ions**, H^+ which are responsible for the reactions of an acid.
- An **alkali solution** contains **hydroxide ions**, OH^-.
- When an acid **neutralises** an alkali, the hydrogen ions react with the hydroxide ions to make water.
  $H^+ + OH^- \longrightarrow H_2O$

The pH scale

- When an acid is added to alkali, or the other way round, a change in pH happens.

adding an alkali to an acid	adding an acid to an alkali
the pH at the start is low	the pH at the start is high
the pH rises as the alkali neutralises the acid when neutral, the pH = 7	the pH falls as the acid neutralises the alkali when neutral, the pH = 7
when more alkali is added, the pH rises above 7	when more acid is added the pH falls below 7

- **Universal indicator solution** can be used to measure the acidity of a solution. A few drops are added to the test solution and then the colour of the solution is compared to a standard colour chart. When acid is added to alkali, they neutralise each other.

Questions

Grades D–C
1. What's the base needed to make zinc sulfate?
2. Which salt is made when magnesium hydroxide reacts with sulfuric acid?

Grades B–A*
3. Write a balanced symbol equation for the reaction between HCl and ZnO.
4. Which ions are responsible for making alkaline solutions?

Reacting masses

Relative formula mass

- **Relative formula masses** need to be added up in the right order if there are brackets in the formula.

- When chemicals react, the atoms of the reactants swap places to make new compounds – the products. These products are made from just the same atoms as before. There are the same number of atoms at the end as there were at the start, so the overall mass stays the same.

$Al(OH)_3$

1 Work out the inside of the bracket first. $16 + 1 = 17$
2 Now multiply the bracket by 3. $17 \times 3 = 51$
3 Work out the outside of the bracket. $= 27$
4 Finally, add them all together. $27 + 51 = 78$

Percentage yield

- Calculations can be made of how much product is produced in a reaction without knowing the equations for the reactions.

- To calculate **percentage yield**, the following two things must be known:
  – the amount of product made, the 'actual yield'
  – the amount of product that should have been made, the 'predicted yield'.

$$\text{percentage yield} = \frac{\text{actual yield}}{\text{predicted yield}} \times 100$$

- Predicted yield can be made by looking at the equation for a reaction.
  $ZnCO_3 \longrightarrow ZnO + CO_2$
  To find out how much CO_2 is made when 625 g of $ZnCO_3$ decomposes, the first step is to find the relative formula mass of $ZnCO_3$.
  $65 + 12 + 16 + 16 + 16 = 125$
  Then find the relative formula mass of CO_2.
  $12 + 16 + 16 = 44$
  The last step is to work out the amount of product made using ratios.
  So if 125 g $ZnCO_3$ gives 44 g CO_2
  then 625 g $ZnCO_3$ gives $\frac{625}{125} \times 44$
  $= 220$ g CO_2

Questions

You may need to use the periodic table on page 4 to help you to answer the questions.

Grades D–C
1 Write down the relative formula mass of calcium nitrate, $Ca(NO_3)_2$.

Grades B–A*
2 Explain why mass is conserved in a chemical reaction.

Grades D–C
3 Tim made 24 g of crystals instead of 32 g. What's the percentage yield?

Grades B–A*
4 How much carbon dioxide is made in the complete thermal decomposition of 59.5 g nickel carbonate, $NiCO_3$?

Fertilisers and crop yield

Using fertilisers

- Farmers use **fertilisers** to increase their **crop yields**. They must first be dissolved in water before they can be absorbed by plants through their roots.

- Fertilisers increase crop yield by replacing the essential elements used by a previous crop. More nitrogen gets incorporated into plant protein, so there's increased growth.

Relative formula mass

- To calculate the yield when making a fertiliser, you need to calculate its **relative formula mass**. The relative formula mass of ammonium nitrate, NH_4NO_3, is 80.

  NH_4NO_3
  14   4 x 1 = 4   14   3 x 16 = 48
  14 + 4 + 14 + 48 = 80

- Farmers can use relative formula masses to find the percentage of each element in a fertiliser – it's printed on the bag.

  $$\text{percentage of element} = \frac{\text{mass of the element in the formula}}{\text{relative formula mass}} \times 100$$

Making fertilisers

- Many fertilisers are **salts**, so they can be made by reacting acids with bases.

  acid + base ⟶ salt + water

  nitric acid + potassium hydroxide ⟶ potassium nitrate + water
  nitric acid + ammonium hydroxide ⟶ ammonium nitrate + water
  sulfuric acid + ammonium hydroxide ⟶ ammonium sulfate + water
  phosphoric acid + ammonium hydroxide ⟶ ammonium phosphate + water

- In the laboratory, sulfuric acid (acid) is reacted with ammonium hydroxide (base). The amounts used in the reaction must be exactly right, so a **titration** is carried out before mixing the main batch of chemicals.
  – Titrate the alkali with the acid, using an **indicator**.
  – Repeat the titration until three consistent results are obtained.
  – Use the titration result to mix the correct amounts of acid and alkali, without the indicator.
  – The fertiliser made is dissolved in water, so evaporate most of the water using a hot water bath.
  – Leave the remaining solution to crystallise, then filter off the crystals.

The dangers of fertilisers

- Fertilisers must be applied carefully. Rain water dissolves fertilisers, which run off into nearby water courses. If nitrate or phosphate levels in a water course are too high, **eutrophication** occurs.

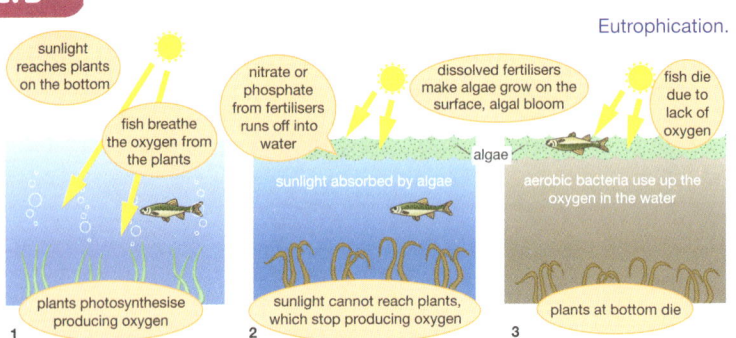

Eutrophication.

1. sunlight reaches plants on the bottom; fish breathe the oxygen from the plants; plants photosynthesise producing oxygen
2. nitrate or phosphate from fertilisers runs off into water; dissolved fertilisers make algae grow on the surface, algal bloom; sunlight absorbed by algae; sunlight cannot reach plants, which stop producing oxygen
3. fish die due to lack of oxygen; aerobic bacteria use up the oxygen in the water; plants at bottom die

Questions

Grades D-C
1 What's the relative formula mass of ammonium phosphate, $(NH_4)_3PO_4$?

Grades B-A*
2 Calculate the percentage of nitrogen in ammonium phosphate.

Grades D-C
3 Which acid and base react to make potassium phosphate?

Grades B-A*
4 Suggest how a solid sample of potassium phosphate could be made. Outline all the main stages.

The Haber process

The Haber process

- The Haber process uses:
  - an iron catalyst
  - high pressure
  - a temperature of 450 °C
  - a recycling system for unreacted nitrogen and hydrogen.

- The word equation for the Haber process is:
  nitrogen + hydrogen ⇌ ammonia

- The symbol equation for the Haber process is:
  $N_2 + 3H_2 \rightleftharpoons 2NH_3$

- As the reaction is reversible, the **percentage yield** for the reaction can't be 100%.
  - A higher pressure **increases** the percentage yield but high pressures costs more.
  - The high temperature **decreases** the percentage yield. However, higher temperatures do make the reaction go faster.
  - 450 °C is an **optimum temperature** – the yield isn't as good, but that yield is made faster, so a satisfactory amount is produced in the right time.
  - Catalysts don't affect the yield – they just make the reaction go faster.

The Haber process.

The costs of ammonia production

- Different factors affect the cost of making a new substance:
  - **labour** – chemical plants are heavily automated and need few people to operate them
  - **reactants** – hydrogen is made from natural gas or by cracking oil, which costs money; nitrogen has to be cleaned, dried and compressed
  - **recycling** of unreacted materials – means that money isn't wasted
  - **high pressure** – this makes the reaction work better but costs more
  - **energy** – the higher the temperature, the more fuel is needed
  - **reaction rate** – the faster the reaction, the more product is made from the same equipment, so the cheaper it is
  - **pollution control** – reducing pollution is expensive.

- Chemical plants work at the conditions that produce the highest percentage yield for a reaction, most cheaply. These are known as the **optimum conditions**.
  - High temperature means higher rates but lower yields, so it runs at the optimum temperature of 450 °C. This temperature means higher energy costs and also lower yields, but the increase in **rate** compensates. The plant produces more ammonia in a day at this temperature than it would at lower temperatures.
  - Total energy costs are not only due to heating costs. The plant needs compressors and pumps to achieve a high pressure. High pressure costs more, so a lower, optimum pressure is used.
  - Although the reaction has a low percentage yield, the unreacted chemicals are recycled, and can go back into the reaction vessel, saving costs.

Questions

Grades D–C
1 At what temperature is the Haber process carried out?

Grades B–A*
2 Why is this temperature chosen?

Grades D–C
3 Why is a low pressure not used in the Haber process?

Grades B–A*
4 High pressure gives the highest yields in the Haber process, but it's not used. Explain why.

Detergents

C4 CHEMICAL ECONOMICS

Detergents

Grades D–C

- A detergent can be made by **neutralising** an organic acid using an alkali.
  acid + alkali ⟶ salt + water
  It's suitable for cleaning uses because:
  – it dissolves grease stains   – it dissolves in water at the same time.

- New washing powders allow clothes to be washed at low temperatures. It's good for the environment to wash clothes at 40 °C instead of at high temperatures because washing machines have to heat up a lot of water. This needs energy, so the lower the temperature of the water, the less energy is used and the less greenhouse gases are put into the atmosphere.

- Washing clothes at low temperatures is also good for coloured clothes as many dyes are easily damaged by high temperatures. It also means that many more fabrics can be machine washed as their structure would be damaged at higher temperatures.

Grades B–A*

- A detergent molecule is made of two parts:
  – the **hydrophilic** part – forms bonds with the water and pulls the grease off the fabric or dish
  – the **hydrophobic** part – forms bonds with the oil or grease.

  hydrophobic end, dissolves in grease
  hydrophilic end, dissolves in water

  A detergent molecule.

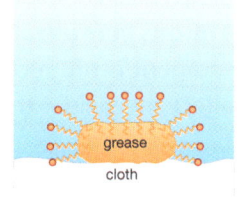

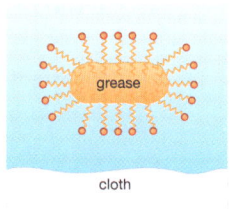

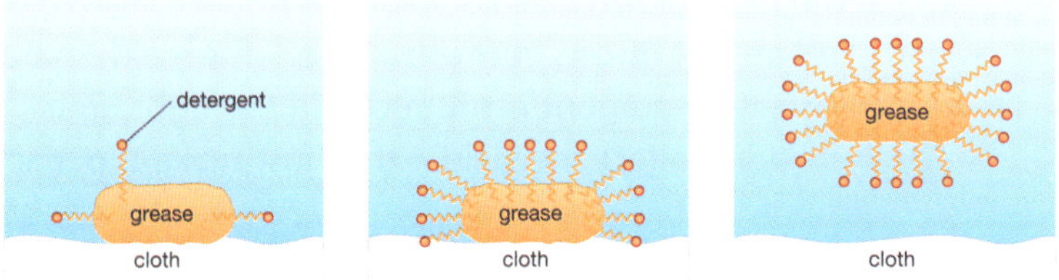

  How a detergent works.

Dry cleaning

Grades D–C

- Some fabrics will be damaged if they are washed in water, so they must be **dry-cleaned**. A dry-cleaning machine washes clothes in an organic solvent. The word 'dry' doesn't mean that no liquids are used, just that the liquid solvent isn't water.

- Most of the stains on clothing contain grease from the skin or from food. Grease-based stains won't dissolve in water, but they will dissolve easily in a dry-cleaning solvent.

Grades B–A*

- Forces between molecules are called **intermolecular molecules**.
  – These forces hold molecules of grease together and molecules of dry-cleaning solvent together.
  – The forces join anything to anything, so dry-cleaning solvent molecules also bind to grease. The grease then dissolves in the solvent.
  – Molecules of water are held together by stronger intermolecular forces called **hydrogen bonds**. The water molecules can't stick to the grease because they're sticking to each other much too strongly.

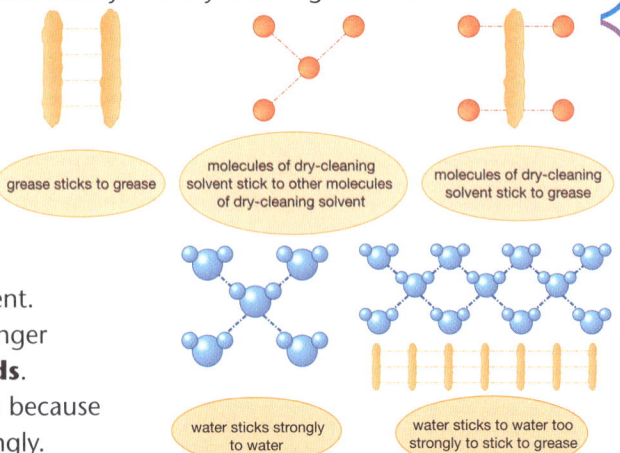

  grease sticks to grease
  molecules of dry-cleaning solvent stick to other molecules of dry-cleaning solvent
  molecules of dry-cleaning solvent stick to grease
  water sticks strongly to water
  water sticks to water too strongly to stick to grease

Questions

Grades D–C
1. What are the two reactants used in making a detergent?

Grades B–A*
2. What does the hydrophobic tail of the detergent dissolve?

Grades D–C
3. Give two reasons why a dry-cleaning method may be used.

Grades B–A*
4. Which intermolecular forces are stronger: grease to water or water to water?

Batch or continuous?

Batch and continuous manufacturing

- Drugs companies make medicines in small batches, which are then stored.
  - New batches are made when the stored medicine runs low. If a lot of one medicine is needed, several batches can be made at the same time.
  - Once they have made a batch of one drug, it's easy to switch to making a different drug.
- The large scale production of **ammonia** is different to the small scale production of pharmaceuticals as it's a continuous process.

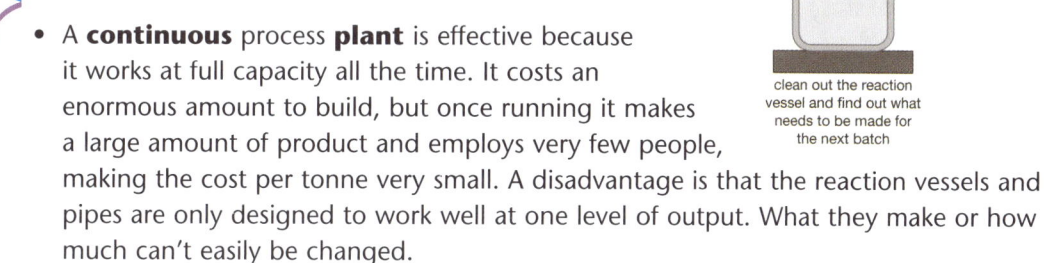

BATCH PROCESS
- put reagents into the reaction vessel
- carry out the reaction
- filter off the product
- clean out the reaction vessel and find out what needs to be made for the next batch

- A **continuous** process **plant** is effective because it works at full capacity all the time. It costs an enormous amount to build, but once running it makes a large amount of product and employs very few people, making the cost per tonne very small. A disadvantage is that the reaction vessels and pipes are only designed to work well at one level of output. What they make or how much can't easily be changed.
- **Batch processes**, by comparison, are flexible. It's easy to change from making one compound to another. Each batch has to be supervised, so labour costs are higher. Also, time spent filling and emptying reaction vessels means the vessels aren't producing chemicals, so they're not used as efficiently as in a continuous process.

Why are medicines so expensive?

- The high costs of making and developing medicines and pharmaceutical drugs include:
  - **strict safety laws**
  - **research and development** – many take years to develop
  - **raw materials** – may be rare and costly
  - **labour intensive** – because medicines are made by a batch process, less automation can be used.

Drug development

- Whether or not a drug is developed depends on a number of economic considerations.
  - Drug development is expensive because it takes time and labour costs are high.
  - Testing drugs to meet legal requirements can take years.
  - There must be enough demand for the drug.
  - The length of time taken to pay back the initial investment is long.

Questions

Grades D-C
1 What's a batch process?

Grades B-A*
2 Write down one advantage of using a continuous process.

Grades D-C
3 Give two reasons why medicines are expensive to develop.

Grades B-A*
4 How long can it take to research, develop, test and patent a new drug?

Nanochemistry

C4 CHEMICAL ECONOMICS

Forms of carbon

Grades D–C

- There are three forms of **carbon** shown in the table.

	diamond	graphite	buckminster fullerene
use	cutting tools – very hard jewellery – lustrous and colourless	pencil leads – slippery lubricants – slippery electrode – conducts electricity and has high melting point	semiconductors in electrical circuits (nanotubes) industrial catalysts reinforce graphite in tennis rackets (nanotubes)
structure			
properties	doesn't conduct electricity because it has no free electrons hard with a high melting point because of the presence of many strong covalent bonds	conducts electricity because of delocalised electrons that can move slippery because layers of carbon atoms are weakly held together and can slide easily over each other high melting point because there are many strong covalent bonds to break	nanotubes have a large surface area so they can be used as cages to trap or transport other molecules or as carriers for catalysts

Grades B–A*

Allotropes

- Different forms of the same element are called **allotropes**. Diamond, graphite and the fullerenes are all **allotropes of carbon**.

- The different arrangements of atoms give each allotrope different properties. In diamond, each atom is held by **covalent bonds** to four other atoms, **tetrahedrally**, which are bonded further in different directions. This is called a **giant structure**. So many strong covalent bonds make the diamond hard and very difficult to melt. The bonding results in no free electrons, so it doesn't conduct electricity.

Grades B–A*

Nano properties

Grades D–C

- A **fullerene** changes at the **nanoscale**. The shape of the individual particles – balls or tubes, sieves or cages – is their nanostructure. This gives them their **nano properties**. These are different from **bulk properties**, which are the properties of large amounts of a material.

- Nanoparticles are made in a different way. One type of nanoparticle is used and then bits are knocked off or stuck on. Structural engineering happens at a molecular level.

Grades B–A*

Questions

Grades D-C
1. Which properties of graphite make it useful as an electrode?
2. Write down two potential uses of nanotubes.

Grades B-A*
3. Explain why graphite can conduct electricity but diamond can't.
4. Explain one use of a fullerene as a cage.

How pure is our water?

Water purification

- The water in a river is cloudy and often not fit to drink. It may also contain pollutants such as nitrates from fertiliser run off, lead compounds from lead pipes, and pesticides from spraying near to water resources. To make clean drinking water it's passed through a **water purification** works.

- There are three main stages in water purification:
  – **sedimentation** of particles – larger bits drop to the bottom
  – **filtration** of very fine particles – sand is used to filter out finer particles
  – **chlorination** – kills microbes.

A sedimentation tank at a water treatment works.

- Some soluble substances remain in the water. Some of these can be poisonous, for example pesticides and nitrates. Extra processes are needed to treat these.

- Seawater has many substances dissolved in it so it's undrinkable. Techniques such as **distillation** must be used to remove the dissolved substances. Distillation uses huge amounts of energy, and is very expensive. It's only used when there's no fresh water.

The importance of clean water

- Clean water saves more lives than medicines. That's why, after disasters and in developing countries, relief organisations concentrate on providing clean water supplies.

- Water is a **renewable resource**, but that doesn't mean the supply is endless. If there isn't enough rain in the winter, reservoirs don't fill up properly for the rest of the year.

- Producing tap water does incur costs. It takes energy to pump and to purify it – all of which increases climate change.

Water tests

- In a precipitation reaction, two solutions react to form a solid that doesn't dissolve.

  lead nitrate + sodium **sulphate** ⟶ lead sulphate (**white** precipitate) + sodium nitrate
  silver nitrate + sodium **chloride** ⟶ silver chloride (**white** precipitate) + sodium nitrate
  silver nitrate + sodium **bromide** ⟶ silver bromide (**cream** precipitate) + sodium nitrate
  silver nitrate + sodium **iodide** ⟶ silver iodide (**yellow** precipitate) + sodium nitrate

- The **balanced symbol equations** for the precipitation reactions above are:

  $Pb(NO_3)_{2(aq)} + Na_2SO_{4(aq)} \longrightarrow PbSO_{4(s)} + 2NaNO_{3(aq)}$
  $AgNO_{3(aq)} + NaCl_{(aq)} \longrightarrow AgCl_{(s)} + NaNO_{3(aq)}$
  $AgNO_{3(aq)} + NaBr_{(aq)} \longrightarrow AgBr_{(s)} + NaNO_{3(aq)}$
  $AgNO_{3(aq)} + NaI_{(aq)} \longrightarrow AgI_{(s)} + NaNO_{3(aq)}$

Questions

Grades D-C
1 Explain why filtration is used in the water purification process.

Grades B-A*
2 Explain why distillation uses large amounts of energy.

Grades D-C
3 What type of reaction takes place between barium chloride and sulfates?

Grades B-A*
4 Write a balanced symbol equation for the reaction between silver nitrate solution and magnesium chloride, $MgBr_2$.

C4 CHEMICAL ECONOMICS

C4 Summary

C4 CHEMICAL ECONOMICS

Chemical industry

- Nitrogen and hydrogen make ammonia in the **Haber process**.
  $N_2 + 3H_2 \rightleftharpoons 2NH_3$

- **Ammonia** reacts as a **base** to form fertilisers such as ammonium nitrate from nitric acid, and ammonium phosphate from phosphoric acid.

- Ammonia is made all the time in a **continuous process**. Pharmaceutical drugs are made on a smaller scale by a **batch process**.

- We can change the **conditions** in the Haber process to give us the best **yield**.
  – High pressure increases yield.
  – High temperature decreases yield but increases rate.
  – Optimum temperatures are used.

- If we know the mass of the reactants, we can work out what mass of products to expect.

- Different factors affect the **cost** of making fertilisers or any new substance. High pressures mean higher energy costs.

- Industry makes chemicals such as **fertilisers**. They must be cheap enough to use.

- We can measure the **percentage yield** of a reaction.
  $\% \text{ yield} = \dfrac{\text{actual yield}}{\text{predicted yield}} \times 100$

- **Water resources** are found in lakes, rivers, aquifers and reservoirs. Water needs testing and purifying before use. Purification includes:
  – filtration
  – sedimentation
  – chlorination.

- **Fertilisers** make crops grow bigger as they provide plants with extra nitrogen, phosphorus and potassium. These are essential chemical elements for plant growth.

- **Detergents** are **molecules** that combine with both grease and water. They have a hydrophilic head and a hydrophobic tail. Washing powders contain detergents but some also contain enzymes, which allow clothes to be washed at a lower temperature. This saves energy.

Nanochemistry

- **Diamonds** are used in cutting tools and jewellery. They're **very hard** and have a **high melting point**.

- **Fullerenes** were discovered fairly recently. Buckminster fullerene has the formula C_{60}.

- The element carbon can exist in different forms. This is due to differences at the **nanoscale**.

- **Graphite** is carbon. It's **slippery** so it can be used as a lubricant. It also **conducts electricity** so it can be used as electrodes.

Moles and empirical formulae

Molar mass

- The mass of one mole of a substance is called its **molar mass**.
- The molar mass is the relative formula mass of a substance in grams.
  The **molar mass** of ammonium sulfate, $NH_4(SO_4)_2$ is $14 + (4 \times 1) + 2(32 + 64) = 210$ g.
  The relative atomic masses: H = 1, O = 16, N = 14, S = 32.

The 'relative' in relative atomic mass

- The relative atomic mass is the mass compared, 'related', to a standard atom.
- We compare everything to a twelfth of carbon-12.
- Relative atomic masses are ratios, so they never have units. The relative atomic mass of an element is the average mass of an atom of the element compared to the mass of an atom of carbon-12.
- Number of moles = $\dfrac{\text{mass of chemical}}{\text{molar mass}}$
- We can use moles to predict masses from equations.

  This is the equation for burning heptane, C_7H_{16}.

  $$C_7H_{16} + 11O_2 \longrightarrow 7CO_2 + 8H_2O$$

  molar mass of C_7H_{16} is $(7 \times 12) + (16 \times 1) = 100$
  molar mass of CO_2 is $(1 \times 12) + (2 \times 16) = 44$

  If 1 mole (100 g) of C_7H_{16} is burned then 7 moles (7×44 g) of CO_2 is made.
  If 100 g of C_7H_{16} is burned then $7 \times 44 = 308$ g of CO_2 is made.

Empirical formulae

- An **empirical formula** tells us the ratio of each type of atom in a compound, but a full formula is also needed.
- The empirical formula is the ratio of each type of atom present in the compound.
- An empirical formula must be a whole number ratio.

  The empirical formula of heptene, C_7H_{14} is the ratio of the atoms, CH_2.

  The empirical formula for glucose, $C_6H_{12}O_6$ and ethanoic acid, CH_3COOH is the same. It is CH_2O for both, yet the two compounds are very different.

- With the masses of each element of a compound, we can work out empirical formulas.
- Stage 1 Write down the mass of each element present.
- Stage 2 Look up the relative atomic mass of each element present
- Stage 3 Work out how many moles of each element present.
- Stage 4 Choose the element present in the lowest amount.
- Stage 5 Divide the moles of each element by the moles in the lowest amount.

2.45 g of sulfuric acid contains 0.025 g H, 0.8 S, 1.6 g O.

	H	S	O
Stage 3 Convert to moles.	0.05	0.8	1.6
	1	32	16
	0.05 mol	0.025 mol	0.1 mol

Stage 4 Element in the smallest amount is sulfur (0.025 mol).

Stage 5 Divide by the lowest.	0.05	0.025	0.1
	÷0.025	÷0.025	÷0.025
	2	1	4

Empirical formula is H_2SO_4.

Questions

Grades D–C

1. What is the molar mass of this acid: $CH_3C(CH_3)_2COOH$?

Grades B–A*

2. How much carbon dioxide is made when 440 g propane, C_3H_8, is burned?

Grades D–C

3. What is the empirical formula for hexane, C_6H_{14}?

Grades B–A*

4. 100 g of sulfuric acid contains 2.04% H, 32.65% S and 65.31% O. Work out the empirical formula of sulfuric acid. Relative atomic masses: H = 1, S = 32, O = 16.

Electrolysis

Decomposing electrolytes

- Electric current is a flow of charge.
- Electrolytes are ionic. The charge moves through the electrolyte by the **ions** moving.
- If the electrolyte solidifies, then the ions cannot move and the **current** cannot flow.
- Positive ions move toward the cathode. Negative ions move toward the anode.
- Ions reach the electrodes and are discharged; they turn into atoms or molecules.
- If the electrolyte is a solution, water is present as well. Hydrogen is made at the cathode and oxygen at the anode. For example: potassium nitrate solution, $KNO_{3(aq)}$.

What happens when a molten electrolyte decomposes?

- The power supply drives electrons round the circuit.
- The cathode pushes electrons onto positive ions.
- The anode pulls electrons off the negative ions. In each case the ions are discharged.

electrolyte	half equation at cathode	half equation at anode
KCl	$2K^+ + 2e^- \rightarrow K$	$2Cl^- - 2e^- \rightarrow Cl_2$
$PbBr_2$	$Pb^{2+} + 2e^- \rightarrow Pb$	$2Br^- - 2e^- \rightarrow Br_2$
Al_2O_3	$2Al^{3+} + 6e^- \rightarrow 2Al$	$6O^{2-} - 6e^- \rightarrow 3O_2$

- Many dissolved compounds decompose just as normal. Some do not. This is because the water also forms ions.
  $H_2O_{(l)} \longrightarrow H^+_{(aq)} + OH^-_{(aq)}$
- If the positive ions are hard to discharge, then H^+ ions from water form hydrogen gas at the cathode: $2H^+ + 2e^- \longrightarrow H_2$
- If the negative ions are hard to discharge, then OH^- ions from water form oxygen gas at the anode: $4OH^- - 4e^- \longrightarrow 2H_2O + O_2$

  $KNO_{3(aq)}$ contains $K^+_{(aq)}$ and $NO_3^-{}_{(aq)}$ ions.
  These ions are hard to discharge, so the water discharges as hydrogen and oxygen.

What affects the amount produced?

- The amount produced at each electrode increases with current and with time.
- The number of ions discharged at the electrodes is affected by the amount of charge transferred.
- The electrolysis of copper(II) sulfate using copper electrodes is a different reaction. The cathode gets heavier as it is plated. The anode gets lighter as it dissolves. They do it by the same amount.

- We can work out how much charge has been transferred by using $Q = It$.
- In the electrolysis of copper(II) sulfate using copper electrodes, the cathode pushes electrons onto copper ions, discharging them. $Cu^{2+} + 2e^- \rightarrow Cu$

time	t	second
current	I	amp
charge	Q	coulomb

- Instead of the anode pulling electrons off ions in the electrolyte, it pulls electrons off the copper anode itself and the copper atoms turn into ions. $Cu - 2e^- \rightarrow Cu^{2+}$
- The mass change of each electrode is the same.

Questions

(Grades D–C)
1 Which type of ions move towards the cathode during electrolysis?

(Grades B–A*)
2 Write down two half equations that represent the electrolysis of PbI_2.

(Grades D–C)
3 Write down two quantities that change the amount produced at an electrode.

(Grades B–A*)
4 How much charge is transferred by a current of 0.2 A for 3 hours?

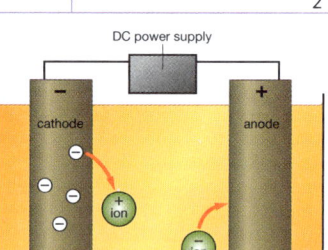

Quantitative analysis

C5 HOW MUCH?

Solute and solutions

Grades D–C

- The substance dissolved in a liquid is the **solute**.
- The more concentrated the solution, the more crowded the solute particles.
- The more dilute the solution, the less crowded the solute particles.

Recommended daily allowances

Grades B–A*

- Sodium ions are essential in diet.
- Too much sodium causes high blood pressure and heart disease.
- The main source of sodium ions is salt.
- Food labels need to be read with care. Some give the amount of sodium, others the amount of salt.

  To convert grams of sodium to grams of salt:
  sodium chloride is NaCl.

  Relative formula mass of NaCl = 23 + 35.5 = 58.5.

  There are 23 g of sodium in 58.5g of salt.

  There is 1 g of sodium in $\frac{58.5}{23}$ = 2.5 g of salt.

  So mass of salt = 2.5 times mass of sodium.

- Prepared foods usually contain other sodium compounds in small quantities, so their sodium ions must be counted as well.

How to dilute solutions

Grades D–C

- If a solution is three times more dilute, it takes up three times more volume.

  You have to make a solution 10 times more dilute.

  You start with 1 cm^3.

  Starting volume is 1 cm^3, final volume is 10 cm^3.

  Extra to be added 10 − 1 = 9 cm^3 of water.

Top Tip!
1 dm^3 is the same as 1 litre. There are 1000 cm^3 in 1 dm^3.

Concentration

Grades B–A*

- If one solution reacts with another, we need to know how many moles of each chemical are dissolved rather than their mass.
- We often measure concentration in mol/dm^3 (mol per dm^3) of solution.

  Concentration = $\frac{\text{number of moles of solute}}{\text{volume of solution in dm}^3}$

Grades D–C

1 What is a solute?

Grades B–A*

2 How many grams of sodium are there in 25 g of salt?

Grades D–C

3 You have a 10 cm^3 of a solution. You want to make it 10 times as dilute. How much water do you add to it?

Grades B–A*

4 There is 1 mole of salt in 2 dm^3 of water. What is the concentration of the solution in mol/dm^3?

Titrations

C5 HOW MUCH?

Titrations

- When an acid reacts with an alkali, the pH of the solution changes (see diagram).
- The point where the acid has just reacted with all the alkali is the end point.
- At the end point the pH changes very suddenly.
- A salt is made.
  acid + alkali ⟶ salt + water
- We use titrations to find the concentration of an alkali from the concentration of an acid. You can learn to calculate the concentration in mol/dm^3 on page 42.
- Three equations have a relationship to each other:
  concentration = number of moles ÷ volume in dm^3
  number of moles = concentration × volume in dm^3
  volume in dm^3 = number of moles ÷ concentration

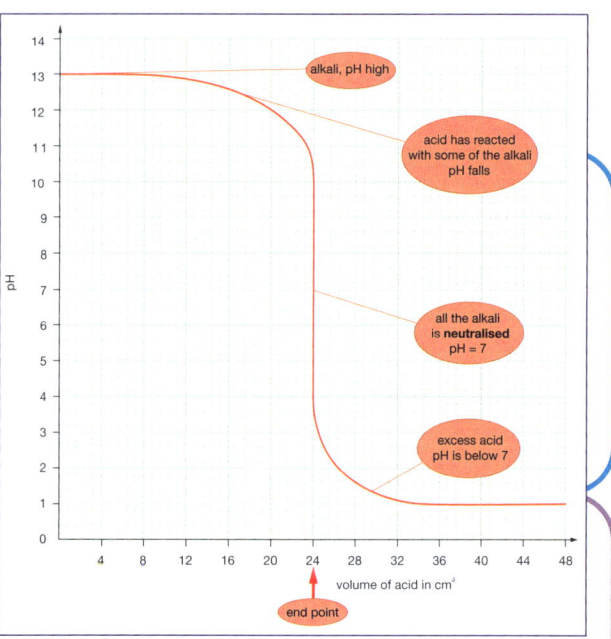

How pH changes when acid is added to alkali.

Calculations

- Indicators, such as phenolphthalein, screened methyl orange and litmus, give a sudden colour change at the neutral point, making the end point in titrations very easy to spot.
- Mixed indicators are not used in titrations, e.g. Universal indicator gives a continuous colour change so it is hard to see the end point..
- Titration is an accurate technique. Large differences in readings show something is wrong. Close readings show the technique is **reliable**. Small differences can be experimental error. Using the average reading allows for this.

If 24.2 cm^3 of a solution of 0.11 mol/dm^3 hydrochloric acid reacts with 25.0 cm^3 of sodium hydroxide solution, what is the concentration of the sodium hydroxide solution?

1. Find how many moles of acid would neutralise one mole of alkali using the equation
   HCl + NaOH ⟶ NaCl + H$_2$O
   – 1 mole hydrochloric acid = 1 mole sodium hydroxide (this is a 1:1 ratio).
2. Find how many moles of acid were used in the titration:
   – number of moles acid = acid concentration × volume in dm^3
   – number of moles acid = 0.11 × 24.2 ÷ 1000 = 0.00266 moles.
3. Find how many moles of alkali were used.
   – number of moles alkali = alkali concentration × volume in dm^3
   – number of moles alkali = alkali concentration × 25.0 ÷ 1000.
4. Link the number of moles of acid with the number of moles of alkali:
   – number of moles acid = number of moles alkali
          0.00266 = alkali concentration × 0.0250
          0.00266 ÷ 0.0250 = alkali concentration = 0.106 mol/dm^3.

Questions

Grades D-C
1. Alkali is slowly added to acid. How does the pH change?

Grades B-A*
2. If number of moles = concentration × volume in dm^3, write down the relationship for concentration.

Grades D-C
3. Why are titrations carried out several times before a volume is decided?

Grades B-A*
4. 23.6 cm^3 of a solution of 0.12 mol/dm^3 hydrochloric acid reacts with 25.0 cm^3 of sodium hydroxide solution. What is the concentration of the sodium hydroxide solution?

Gas volumes

Measuring the gas made in a reaction

- An upturned burette or cylinder, needs to be filled with water before you turn it. The volume is read off the scale on the side.
- The scale on a burette goes the opposite way to that on a measuring cylinder.
- When using a balance, a loose plug of cotton wool is put in the neck of the flask. This lets the gas out but not any spray of liquid droplets.
- You can tell the number of moles of gas just by knowing its volume.
- Under normal conditions (room temperature and pressure), 1 mole of particles of any gas takes up 24 dm^3.
- All that matters is the number of particles. It doesn't matter how big they are. This is because the space between molecules in a gas is enormous. The actual size of each molecule is unimportant.

$$\text{number of moles} = \frac{\text{volume of gas in dm}^3}{24}$$

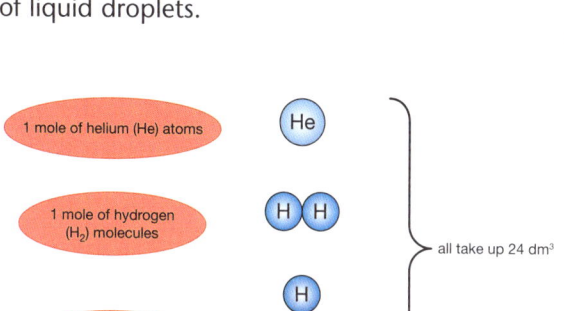

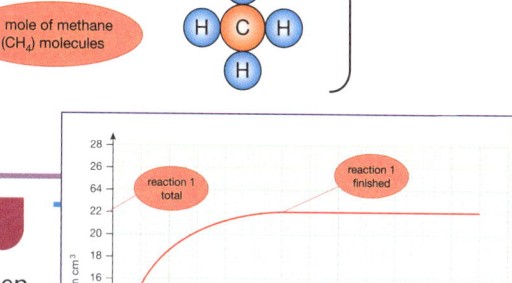

1 mole of helium (He) atoms
1 mole of hydrogen (H$_2$) molecules
1 mole of methane (CH$_4$) molecules
all take up 24 dm^3

Changing the amounts of reactants

- Magnesium ribbon reacts with acid to give off hydrogen gas. If you use exactly the right amounts, no magnesium ribbon is left at the end, and all the acid is neutralised.
- If you use the same amount of acid but only half the amount of magnesium, you will only get half the gas. In this case we say the magnesium is the **limiting reactant**.
- The total amount of gas produced is directly proportional to the amount of the limiting reactant.

Graph A

What can you tell from the results?

- Remember to look at the angle of the curve to see what is happening to the **rate of the reaction**. The steeper the curve, the faster the reaction.
- You can find rate of reaction at any point by finding the gradient of the curve at that point.

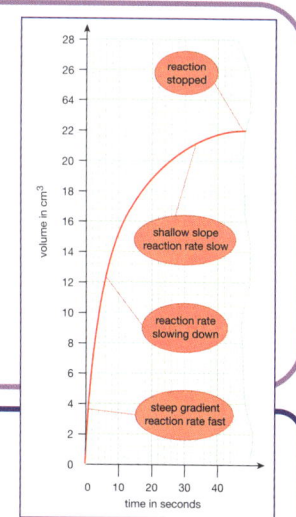

Graph B

Questions

Grades D-C

1 What is the volume of gas collected in this gas syringe?

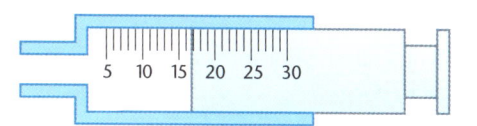

Grades B-A*

2 A reaction gave off 96 cm^3 of sulfur dioxide. How many moles were made?

Grades D-C

3 Explain why the two curves in graph A are different.

Grades B-A*

4 What is the rate of reaction in the first 10 seconds in graph B?

Equilibria

Reversible reactions and equilibrium

- At equilibrium:
  - the rate of the forward reaction equals the rate of the backward reaction
  - concentrations of reactants and of products do not change.
- If the concentration of reactants is greater than the concentration of products, we say that the equilibrium position is on the left.
- If the concentration of the reactants is less than the concentration of the products, we say that the equilibrium position is on the right.

Top Tip! If the forward reaction is endothermic, a higher yield is favoured by an increase in temperature.

- The equilibrium system, reactant ⇌ product, always reduces the effect of any changes just made.
- Concentration:
  - add more reactant, equilibrium moves to the right
  - add more product, equilibrium moves to the left.
- Pressure:
  - increase the pressure, equilibrium shifts to the side with fewer gas molecules
  - decrease the pressure equilibrium moves to the side with more molecules.
- Temperature:
  - increase in temperature in an exothermic forward reaction pushes the equilibrium to the left, reducing the temperature increase
  - if the forward reaction takes in heat (endothermic), an increase in temperature pushes the equilibrium to the right, reducing the temperature increase.

Contact Process

- The position of the equilibrium can be changed if you change the concentration of the reactants or the products, the pressure or the temperature.
- The main stages of the Contact Process are:
  1. sulfur + oxygen ⟶ sulfur dioxide
  2. sulfur dioxide + oxygen ⇌ sulfur trioxide (the reversible reaction)
  3. sulfur trioxide + water ⟶ sulfuric acid
- For the most economic yield, the reaction is carried out at around 450 °C, at atmospheric pressure, using a catalyst of vanadium pentoxide, V_2O_5.
- The equations for the three stages of the contact process are:
  1. $S + O_2 \longrightarrow SO_2$
  2. $2SO_2 + O_2 \rightleftharpoons 2SO_3$ (the equilibrium in Stage 2 lies to the right)
  3. $SO_3 + H_2O \longrightarrow H_2SO_4$.
- Conditions used are **450 °C**. This is a compromise. The forward reaction is exothermic, so high temperatures reduce the yield, but high temperatures increase the rate of reaction.
- **Atmospheric pressure**. Another compromise. High pressure increases the yield, however, the equilibrium lies to the right, so the cost of stronger equipment is not worth it.
- **Catalysts** do not affect the position of the equilibrium. However, they do make the reaction go faster, so more is produced every second.

Questions

Grades D-C
1. At equilibrium, the reaction A + B ⇌ C + D has 1 mole of A and B and 3 moles of C and D. Where is the position of the equilibrium?

Grades B-A*
2. What happens to the equilibrium in the reaction of gases: $N_2 + 3H_2 \rightleftharpoons 2NH_3$ if the pressure is increased.

Grades D-C
3. What are the conditions used in the Contact Process?

Grades B-A*
4. Why is it called a compromise temperature in the Contact Process?

Strong and weak acids

Strong and weak acids

Grades D–C

- Acids have hydrogen atoms in their formula, e.g. hydrochloric acid is HCl.
- In water, the acid molecule ionises – it turns into ions.

  acid molecule ⟶ hydrogen ions + other ions

- The reactions of acids are caused by the hydrogen ions.
- Strong acids change **completely** into their ions when they are put into water.

  strong acid ⟶ hydrogen ions + other ions

  There are lots of hydrogen ions, so the acid seems **very acidic**.

- In weak acid only a few of the molecules change into ions when they are put into water. A reversible reaction is set up.

  weak acid ⇌ hydrogen ions + other ions

  The solution contains lots of acid molecules, but not many H^+ ions – so it does not seem to be so acidic.

Top Tip! Don't forget, weak acids do not fully ionise.

Grades B–A*

- Low pH number = high concentration of H^+.
- Higher pH number = lower concentration of H^+.
- Strong acids ionise completely in water. $HCl \longrightarrow H^+ + Cl^-$
- The high concentration of H^+ means that the pH is low.
- Ethanoic acid is $CH_3COO\underline{H}$. The underlined hydrogen is the one that makes it an acid.
- The equation for the reversible reaction is: $CH_3COOH \rightleftharpoons H^+ + CH_3COO^-$

  As this equilibrium lies to the left, the concentration of H^+ is low and the pH is higher.

Reactions with acids

Grades D–C

- Strong and weak acids have the same reactions.
- The reactions of acids are due to the hydrogen ions.
- Some weak acids have so few hydrogen ions that they will only react with very reactive substances such as magnesium.
- Weak acids react more slowly than strong acids.
- Strong acids produce lots of H^+ ions; there are lots collisions between H^+ and the magnesium. Reactions are fast.
- Weak acids have fewer H^+, so fewer collisions. Reactions are slow.

Grades B–A*

- It is not the total number of collisions that matters; it is the number of collisions every second. This is called the **collision frequency**.
- In strong acids, the concentration of H^+ is higher, so collision frequency is greater.
- In weak acids, the H^+ concentration is lower, so collision frequency is lower.

Questions

Grades D–C

1 Strong acids break into their ions completely if put into water. What happens to weak acids when they are put into water?

Grades B–A*

2 Explain why ethanoic acid is a weak acid. Use ideas about equilibrium.

Grades D–C

3 One ion makes a solution into an acid. Which one?

Grades B–A*

4 Explain why strong acids react faster than weak acids. Use ideas about collisions.

Ionic equations

Precipitation reactions

- In solid ionic compounds, such as sodium chloride, the sodium ions and the chloride ions are held together in fixed positions; they cannot move about.
- When the ionic compound is melted or dissolved, the ions can move about.
- In the reaction of lead nitrate with sodium iodide there is only a reaction between lead ions and iodide ions. The sodium nitrate is the left over solution of ions at the end.

  lead nitrate(aq) + sodium iodide(aq) ⟶ lead iodide(s) + sodium nitrate(aq)

- To write equations for precipitation you have to know which compounds are insoluble.
- Insoluble compounds are: silver chloride, silver bromide, silver iodide, barium sulfate.

Spectator ions

- The reaction between lead nitrate and sodium iodide is the reaction of lead ions with iodide ions.
- The sodium ions from the sodium iodide and the nitrate ions from the lead nitrate are still moving around in the solution.
- As they have not taken any part in the reaction, they are called **spectator ions**.
- The precipitate is also made of ions, but they are trapped in the solid, they cannot move.

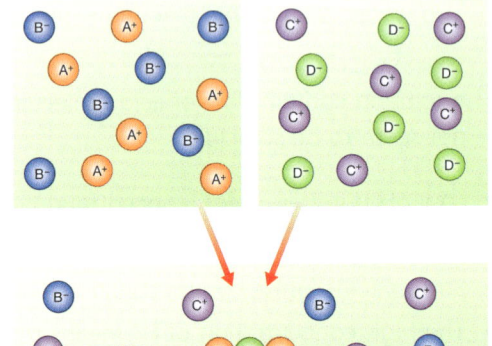

Preparing a clean dry sample of an insoluble salt by precipitation

- **Stage 1** Mix. The precipitate of lead chloride and a solution of sodium nitrate is made.
- **Stage 2** Filter. The precipitate is left with traces of the sodium nitrate solution.
- **Stage 3** Wash. Distilled water removes the traces of sodium nitrate solution.
- **Stage 4** Dry. The precipitate is left in a warm place for the water to evaporate.

Ionic equations

- Word equation:

  lead nitrate$_{(aq)}$ + sodium iodide$_{(aq)}$ ⟶ lead iodide$_{(s)}$ + sodium nitrate$_{(aq)}$

- Symbol equation:

  $Pb(NO_3)_{2(aq)}$ + $2NaI_{(aq)}$ ⟶ $PbI_{2(s)}$ + $2NaNO_{3(aq)}$

- In an **ionic equation** we put the reacting ions only and leave the spectator ions out.

  $Pb^{++}_{(aq)}$ + $2I^-_{(aq)}$ ⟶ $PbI_{2(s)}$

- As the ions are spread completely through each solution and are also moving rapidly, the ions can collide. High collision frequency means that the reaction takes place quickly.

Questions

Grades D-C
1 Write a word equation for the reaction of silver nitrate and sodium iodide.

Grades B-A*
2 Why are sodium ions and nitrate ions often spectator ions?

Grades D-C
3 Why do you wash a precipitate in distilled water when making a dry sample?

Grades B-A*
4 Write an ionic equation for the reaction of silver nitrate and sodium bromide.

C5 Summary

Moles and formulae

The **relative formula mass** of $CH_3C(CH_3)_2CH_3$ is: $(12 \times 5 + (1 \times 12)) = 72$ (relative atomic masses C = 12, H = 1).

The empirical formula of $C_5H_{11}COOH$ is C_3H_6O.

The **relative atomic mass** of an element is the average mass of an atom of the element compared to the 1/12th mass of an atom of carbon-12.

The **empirical formula** is the ratio of each type of atom present in the compound. An empirical formula must be a whole number ratio.

Number of moles = $\dfrac{\text{mass of chemical}}{\text{molar mass}}$.
We can use moles to predic masses from equations.

Electrolysis

In the electrolysis of copper sulfate using copper electrodes, the negative electrode gains mass but the positive electrode loses mass.

Molten $PbBr_2$ **decomposes** in electrolysis to make lead metal and bromine gas.
$Pb^{2+} + 2e^- \rightarrow Pb$
$2Br^- - 2e^- \rightarrow Br_2$

The amount of substance produced at each electrode increases as current increases and as time increases.

Electrolysis of aqueous solutions, such as $K_2SO_{4(aq)}$, often discharge the ions from the water (to give hydrogen and oxygen) rather than ions from the solute.

The charge moves through an electrolyte as the **ions** can move. If the electrolyte is solid then the ions cannot move and the **current** cannot flow.

Adding acid

In a **titration** there is a sudden change in pH at the **end point**.

At the start of a titration, if there is just alkali, the pH number is high. As the acid is added it starts to **neutralise** the alkali, so the pH falls.

When an acid reacts with an alkali the **pH** of the solution changes.

A **strong acid** is one that ionises completely in a solution.
A **weak acid** does not completely ionise and some molecules do not turn into ions.

Equilibria

If the concentration of **reactants** is greater than the concentration of **products**, we say that the **equilibrium position** is on the left.

The Contact Process uses a catalyst of V_2O_5. The conditions used are a temperature of 450 °C and atmospheric temperature.

The three stages of the Contact process are:
1. $S + O_2 \longrightarrow SO_2$
2. $2SO_2 + O_2 \rightleftharpoons 2SO_3$
3. $SO_3 + H_2O \longrightarrow H_2SO_4$

The position of the equilibrium can be changed if you change the concentration of the reactants or the products, the pressure or the temperature.

C5 HOW MUCH?

Energy transfers – fuel cells

Electric current from fuel cells

Top Tip! A fuel cell converts chemical energy directly into electrical energy. There is no heat.

- Fuel cells are used in spacecraft because they:
  - are efficient as they waste very little energy
  - are lighter than normal batteries, so the spacecraft can carry a bigger payload
  - can be used continuously as they don't need time out to be recharged
  - don't need a special fuel with its own separate storage system; the spacecraft has to carry hydrogen and oxygen anyway for the rocket engines
  - produce water that is used by the astronauts for drinking.
- Car makers are very interested in fuel cells as more laws are being passed to reduce the pollution from vehicle exhausts and petrol reserves will eventually run out.
- A normal car engine converts chemical energy into heat, and the heat is then converted to movement energy. This stage is inefficient, and when fuels burn the temperature gets hot enough to react the nitrogen and oxygen in the air to make nitrogen oxides. Nitrogen oxides in the air cause photochemical smog.
- Cars powered by fuel cells are more efficient than normal engines so they use less fuel. The fuel does not burn, so no high temperatures are involved and no oxides of nitrogen are produced. The main product of the hydrogen-powered fuel cell is water, which is not a pollutant at all.
- A problem with cars powered by fuel cells is that hydrogen is a gas. This is more difficult to store inside the car, and filling stations will need a totally different type of fuel pump.

- Fuel cells are especially useful for mobile energy sources.
- The energy is converted directly from chemical energy in the fuel into electrical energy.
- The energy conversion is all done in a single stage.
- This one stage is highly efficient. Almost all the energy is converted.
- If fuel cells replace other fuels, then the main product is water with no nitrogen oxides.
- Fuel cells which replace conventional batteries cause fewer disposal problems.
- They weigh less. Fuel cells would make the cars lighter, reducing fuel consumption.

Reactions of hydrogen and oxygen

- Energy is only released when fuel reacts with oxygen from the air.
- A reaction which gives out energy is exothermic.
- The reaction between hydrogen and oxygen is **exothermic**.
- In a fuel cell that uses hydrogen, the reaction is: hydrogen + oxygen ⟶ water
- If hydrogen reacts with oxygen by burning, the chemical energy is given out as heat.
- A fuel cell converts chemical energy directly into electrical energy – there is no heat loss.
- If the fuel is hydrogen, the reaction is $2H_2 + O_2 \longrightarrow 2H_2O$
- When the fuels react, chemical energy is given out.
- We can show this in an **energy level diagram**.
- An exothermic reaction gives out energy.
- The negative electrode in the fuel cell loses electrons. $2H_2 \longrightarrow 4H^+ + 4e^-$
- The positive electrode in the fuel cell gains electrons. $2H_2O + O_2 + 4e^- \longrightarrow 4OH^-$
- Reactions where electrons are gained and lost are called **redox** reactions.

An energy level diagram.

Questions

Grades D-C
1 Give two reasons why the car industry wants to develop fuel cells.

Grades B-A*
2 Give three advantages of generating electricity with a fuel cell over other ways.

Grades D-C
3 Why is the reaction of hydrogen and oxygen described as exothermic?

Grades B-A*
4 Which electrode reaction loses electrons in the fuel cell?

Redox reactions

Rusting and redox

Grades D–C

- Rusting is a redox reaction.
  iron + oxygen + water ⟶ hydrated iron(III) oxide
- In redox reactions something is oxidised and something else is reduced.
- To stop iron rusting, one of the reactants must be taken out of the equation:
  – water must not touch the iron; iron will not rust in air if the air is very dry
  – oxygen must not touch the iron.
- Covering the iron with a layer of oil, grease or paint stops the iron rusting because they stop oxygen or water from reaching the surface of the iron.

Redox reactions

Grades B–A*

- Oxidation is the addition of oxygen and reduction is the removal of oxygen.
- **O**xidation **i**s when electrons are **l**ost (**Oil**).
- An oxidising agent takes electrons off a substance.
- **R**eduction **i**s when electrons are **g**ained (**Rig**).
- A reducing agent pushes electrons onto a substance.
- When iron rusts:
  – iron loses electrons, it is oxidised
  – oxygen gains electrons, it is reduced.
- Chlorine is like oxygen, it is normally reduced as it gains electrons.

Displacement reactions

Grades D–C

- Word equations for displacement reactions show that the more reactive metal 'swaps places' with the less reactive metal. Magnesium is more reactive than zinc.

  magnesium + zinc sulfate ⟶ magnesium sulfate + zinc
  ↓   ↓
  more reactive metal   less reactive metal

- As the order of reactivity is magnesium, zinc, iron, tin:
  – magnesium metal will displace zinc, iron and tin
  – zinc will displace iron and tin
  – iron will displace tin.

Top Tip!
The more reactive metal becomes ions in solution. The less reactive metal becomes (or stays) as metal.

Grades B–A*

- All metals try to lose electrons to turn into ions.
- The more reactive the metal, the more easily it loses electrons.
- These electrons go on to the ions of other metals which are not so reactive.
- $Mg + ZnSO_4 \longrightarrow MgSO_4 + Zn$ $Mg + Zn^{++} \longrightarrow Mg^{++} + Zn$
  $Zn + FeCl_2 \longrightarrow ZnCl_2 + Fe$ $Zn + Fe^{++} \longrightarrow Zn^{++} + Fe$
- Iron can be protected by coating it with another metal.
- Iron coated with a thin layer of zinc is '**galvanised** iron'. When the zinc layer is scratched the iron does not rust. Zinc is the more reactive, so it gives off electrons instead of the iron. If the iron cannot give off electrons, it cannot react and so it does not corrode. The zinc has sacrificed itself, this is **sacrificial protection**.

Questions

Grades D–C
1. Why does painting prevent rust?

Grades B–A*
2. What happens to electrons in the process of oxidation?

Grades D–C
3. Write a word equation for the reaction between zinc and iron nitrate.

Grades B–A*
4. Write a balanced equation for the reaction between magnesium and iron sulfate ($FeSO_4$).

Alcohols

Fermentation

- The word equation for fermentation is:

  glucose ⟶ ethanol + carbon dioxide

- An optimum temperature between 25 and 50 °C is needed for fermentation.
- The ethanol produced is dilute. The solution is distilled get more ethanol.
- The **molecular formula** for ethanol is C_2H_6O (often written as C_2H_5OH).
- The **displayed formula** for ethanol is:

  $H-\underset{\underset{H}{|}}{\overset{\overset{H}{|}}{C}}-\underset{\underset{H}{|}}{\overset{\overset{H}{|}}{C}}-O-H$

- The balanced chemical equation for fermentation is:

  $C_6H_{12}O_6$ ⟶ $2C_2H_5OH + 2CO_2$
  glucose   ethanol

- Fermentation is carried out under carefully controlled conditions:
  - if the temperature is too cold, the enzymes in yeast will be inactive
  - if the temperature is too hot, the enzymes in yeast will be denatured
  - if air is present, ethanoic acid will be produced instead of ethanol.
- There is a series of alcohols. They have the general formula $C_nH_{2n+1}OH$.

	Molecular formula	Displayed formula
methanol	CH_3OH	$H-\overset{\overset{H}{|}}{\underset{\underset{H}{|}}{C}}-O-H$
ethanol	C_2H_5OH	$H-\overset{\overset{H}{|}}{\underset{\underset{H}{|}}{C}}-\overset{\overset{H}{|}}{\underset{\underset{H}{|}}{C}}-O-H$
propanol	C_3H_7OH	$H-\overset{\overset{H}{|}}{\underset{\underset{H}{|}}{C}}-\overset{\overset{H}{|}}{\underset{\underset{H}{|}}{C}}-\overset{\overset{H}{|}}{\underset{\underset{H}{|}}{C}}-O-H$
butanol	C_4H_9OH	$H-\overset{\overset{H}{|}}{\underset{\underset{H}{|}}{C}}-\overset{\overset{H}{|}}{\underset{\underset{H}{|}}{C}}-\overset{\overset{H}{|}}{\underset{\underset{H}{|}}{C}}-\overset{\overset{H}{|}}{\underset{\underset{H}{|}}{C}}-O-H$

Ethanol from ethene

- Ethanol which is used industrially can be made from ethene by a hydration reaction.

  ethene + water ⟶ ethanol

- Ethene and steam are passed over a hot phosphoric acid catalyst.
- To dehydrate the ethanol, the vapour is passed over a hot aluminium oxide catalyst. ethanol ⟶ ethene + water
- Ethanol can be made by the hydration of ethane. $C_2H_4 + H_2O$ ⟶ C_2H_5OH
- Ethanol can be dehydrated to produce ethene. C_2H_5OH ⟶ $C_2H_4 + H_2O$
- Ethanol can be made in two ways: by fermentation or the hydration of ethene.
- Fermentation:
  - advantages: it produces ethanol from renewable resources
  - disadvantages: large areas of natural forest are cut down to make room for the crops.
- In the UK, industrial ethanol is made from ethene produced from oil and natural gas:
  - advantages: much cheaper; the UK climate is poor for crops for fermentation
  - disadvantages: it is not a renewable method; once the oil has run out; an alternative source of ethene will have to be found.

Questions

Grades D-C

1 What is the word equation for the production of ethanol from sugars?

Grades B-A*

2 Explain what happens if the temperature is too hot during fermentation.

Grades D-C

3 What is the catalyst needed for the dehydration of ethanol?

Grades B-A*

4 Write down the balanced symbol equation for the hydration of ethene.

Chemistry of sodium chloride (NaCl)

Salt

- Rock formations with salt deposits exist in the UK. Cheshire has the main salt mines.
- The salt may be extracted from the rock by using underground cutting machines. This impure salt is called rock salt and is used for gritting roads.
- Salt is also extracted by drilling a borehole down into the salt layer and pumping water down. The salt dissolves and is pumped up. The water is evaporated leaving table salt.
- Once the salt is removed from a mine, the ground may subside, which used to be a major problem. Some houses in these areas had steel frames to protect them if the ground did start to move. Nowadays mining is much more carefully controlled.

Electrolysis of molten sodium chloride

- Molten sodium chloride contains Na^+ ions and Cl^- ions.
- During electrolysis the Na^+ ions move to the negative electrode, the cathode, where they are discharged to form sodium metal. Cathode half reaction: $2Na^+ + 2e^- \longrightarrow 2Na$
- The Cl^- ions move to the positive electrode, the anode. When they are discharged they form chlorine atoms, and then chlorine gas. Anode half reaction: $2Cl^- - 2e^- \longrightarrow Cl_2$

Electrolysis of sodium chloride solution

- Sodium chloride solution is also known as **brine**. When brine is electrolysed, hydrogen is discharged from the water and chlorine from the sodium chloride, leaving sodium hydroxide in solution:

  water + sodium chloride $\longrightarrow$ hydrogen + chlorine + sodium hydroxide

  – the sodium chloride must be a concentrated solution in water
  – the electrodes must be inert
  – hydrogen is produced at the cathode and chlorine is produced at the anode
  – sodium hydroxide solution is formed in the cell.

- Hydrogen and chlorine react explosively so a porous barrier in the middle of the cell allows the ions to move but separates the gases.
- Chlorine is a powerful bleach and a highly poisonous gas. It can be reacted with sodium hydroxide to make a much safer substance which releases chlorine easily when it is needed. Household bleaches are made in this way.

Electrolysis of a brine cell.

- If the salt is dissolved in water, two ions can be liberated at the cathode:
  – Na^+ from the salt and H^+ from the water.
- It is much easier to give electrons to hydrogen ions than to sodium ions, so hydrogen is discharged. Cathode half reaction: $2H^+ + 2e^- \longrightarrow H_2$
- Two ions could be liberated at the anode: Cl^- ions from the salt, and OH^- ions from the water.
- It is easier to discharge Cl^- ions, so chlorine gas is formed.
  Anode half reaction: $2Cl^- - 2e^- \longrightarrow Cl_2$

Questions

Grades D–C
1 What are the different ways that table salt and gritting salt are made?

Grades B–A*
2 Write the equation for the discharge of sodium ions at the cathode.

Grades D–C
3 Why is a barrier used in the cell of the electrolysis of brine?

Grades B–A*
4 Write the equation for the discharge of ions at the anode in the electrolysis of brine.

Depletion of the ozone layer

The ozone layer

- The ozone layer is in the **stratosphere**. There are only tiny amounts of ozone in this layer, but it still absorbs most of the ultraviolet light from the Sun.
- In 1985, scientists discovered that the amount of ozone high over the South Pole was much less than it should be, and they called this the 'hole' in the ozone layer.
- The more depleted the layer of ozone gets, the more ultraviolet radiation can get through to the Earth's surface.
- When CFCs were first discovered, they were thought to be totally safe. We now know that these molecules slowly move up into the stratosphere where they 'attack' the ozone layer.
- CFCs were used as refrigerants and aerosol propellants but they are now banned.

How ozone works

- Visible light can go through the ozone layer very easily, but ultraviolet is absorbed.
- The ultraviolet part of the electromagnetic spectrum has exactly the right frequency to make ozone molecules vibrate.
- The energy of the ultraviolet is converted into movement energy inside each molecule.
- The thicker the ozone layer, the greater the amount of ultraviolet radiation absorbed.
- CFCs do not react in the lower atmosphere, so they are able to diffuse up to the stratosphere which is where they can do damage causing a hole in the ozone layer.
- CFCs do not stay above where they were released. This is why there is a global ban.

CFCs, ozone and free radicals

- In the stratosphere the ultraviolet radiation from the Sun is strong enough to break single chlorine atoms off the CFC molecule.
- A single chlorine atom is called a chlorine **free radical**.
- These chlorine free radicals attack ozone molecules, turning the ozone back into oxygen gas and depleting the ozone layer.
- CFCs are removed from the stratosphere only very slowly, so each CFC molecule has time to do a lot of damage.
- The main alternatives are alkanes and HFCs, hydrofluorocarbons.
- HFCs cannot make chlorine free radicals, so they are safer.

Top Tip! Free radicals are single atoms formed when a covalent bond is split evenly.

- A covalent bond is made of two electrons. When the bond breaks, by UV, the bond can split into equal halves to make free radicals. Free radicals are highly reactive.
- A chain reaction is set off producing another chlorine free radical, which reacts further.
- Chlorine free radicals react with ozone molecules, creating more chlorine free radicals.

    $Cl\bullet + O_3 \rightarrow OCl\bullet + O_2$ $OCl\bullet + O_3 \rightarrow Cl\bullet + 2O_2$

    Combining these two equations gives us $2O_3 \rightarrow 3O_2$

- Eventually two free radicals collide. One possible termination reaction is $Cl\bullet + Cl\bullet \rightarrow Cl_2$
- The inertness of CFCs means that nothing will react with them, so they stay in the stratosphere until they are broken down by ultraviolet light which can take 20–50 years.

Questions

Grades D-C

1. How do the levels of UV light change when there is a depletion of ozone?

Grades B-A*

2. Explain how ozone is able to absorb UV light in the stratosphere.

Grades D-C

3. How does a single chlorine atom cause the ozone to be destroyed?

Grades B-A*

4. Write two equations between chlorine atoms and ozone to show a chain reaction.

Hardness of water

What causes hard water?

Grades D–C

- Permanent hardness is produced when calcium sulfate dissolves in water.
- Temporary hardness is produced when water with dissolved carbon dioxide from the air reacts with calcium carbonate. Together they form calcium hydrogencarbonate.

  calcium carbonate + carbon dioxide + water ⟶ calcium hydrogencarbonate

- A way to compare hardness in samples of water is to measure how much soap reacts. Shake a soap flake with a water sample; the calcium ions in the water will react with the soap and turn it into scum. Shake in more flakes until the soap has reacted with all of the calcium ions. Add more soap to produce a stable lather for the first time. The number of soap flakes used tells you how hard the water is.

Top Tip!

Magnesium and calcium behave the same:

calcium sulfate	magnesium sulfate	calcium carbonate	magnesium carbonate
↓	↓	↓	↓
permanent hardness	permanent hardness	temporary hardness	temporary hardness

Grades B–A*

- Calcium carbonate does not dissolve in pure water, but does in rainwater.
- Rainwater has carbon dioxide dissolved in it from the air, making it slightly acidic.
- A reaction with all three substances produces soluble calcium hydrogencarbonate.

  $CaCO_{3(s)} + CO_{2(g)} + H_2O_{(l)} \longrightarrow Ca(HCO_3)_{2(s)}$
  calcium carbonate   calcium hydrogencarbonate

- Calcium hydrogencarbonate provides the calcium ions in temporary hard water.
- Magnesium carbonate behaves in the same way as calcium carbonate with rain water.

Removing hardness

Grades D–C

- Calcium hydrogencarbonate decomposes on heating to insoluble calcium carbonate, water and carbon dioxide. This makes a solid deposit of limescale inside hot water pipes.
- Permanent hardness is not affected by heating. The calcium sulphate is too stable.
- Ion-exchange resins remove both temporary and permanent hardness. The water flows over solid resin which traps calcium and magnesium ions on to it, taking these ions out of the water.
- Strong acids such as hydrochloric acid remove the limescale, but might react with tap metal.
- Descalers contain weak acids which are less likely to damage anything else.

  acid + carbonate ⟶ salt + carbon dioxide + water

Grades B–A*

- Methods of softening water turn ions that react with soap into an insoluble compound.
- Thermal decomposition softens temporary hardness only. The equation is:

  $Ca(HCO_3)_{2(aq)} \longrightarrow CaCO_{3(s)} + CO_{2(g)} + H_2O_{(l)}$

- Washing soda dissolves putting carbonate ions in solution, making insoluble $CaCO_3$.

  $Ca^{++}_{(aq)} + CO_3^{=}_{(aq)} \longrightarrow CaCO_{3(s)}$

- Ion-exchange columns start with sodium ions on the surface of its resin beads.
- As calcium ions flow past they stick to the resin, pushing sodium ions off into the water.
- The sodium ions in the water do not affect soap so the water is now soft.
- Once the resin surface is coated with calcium ions it is recharged with salt solution.
- The action of acid limescale remover is:

  $2HCl + CaCO_3 \longrightarrow CaCl_2 + CO_2 + H_2O$

Questions

Grades D-C

1. Why does calcium carbonate dissolve in rainwater but not in pure water?

Grades B-A*

2. Write the balanced equation of magnesium carbonate reacting in rainwater.

Grades D-C

3. Which kind of hardness of water do ion-exchange resins remove?

Grades B-A*

4. Explain why concentrated salt solution is later added to ion-exchange resin.

C6 CHEMISTRY OUT THERE

Natural fats and oils

What are fats and oils?

- Fats and oils are compounds called **esters**. They are made of chains of carbon atoms.
- If the carbon atoms in the chain are linked by single bonds, the compound is saturated.
- If the carbon chains contain one or more double bonds, the oil is unsaturated.
- Bromine water is orange. If it is shaken with an unsaturated compound it loses its colour. It is decolourised. This is a test for unsaturation.
- One use of vegetable oils is to make margarine. Vegetable oils are unsaturated.
- The first stage is to 'harden' them and turn them into saturated compounds.
- Hydrogen is bubbled through the oil at about 200 °C using a nickel catalyst.
- The hydrogen reacts with the double bonds and turns them into single bonds.

if all the carbon atoms in a chain are linked by single bonds the compound is saturated

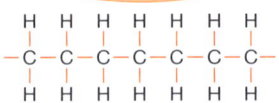

some fats and oils have carbon that contain one or more double bonds, the compounds are unsaturated

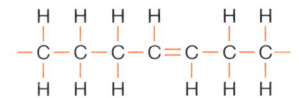

Fats, oils and health

- Saturated fats usually come from animals, unsaturated fats and oils from plants.
- 'Polyunsaturated' means more than one double bond.
- People whose diet is rich in unsaturated oils usually have lower levels of cholesterol.
- Bromine water is a solution of bromine. The bromine molecules make the liquid orange.
- Bromine reacts with the double bonds in the carbon chain in an unsaturated oil.
- The reaction uses up the bromine molecules, so the colour disappears.

Mixing oil and water

- Oil and water are immiscible liquids. They do not mix. They do not dissolve in each other, but it is possible to disperse tiny droplets of one liquid inside the other.
- Water and oil can make two types of emulsion.
- Oil and water can be made to mix as an emulsion by using an emulsifier.
- Fats and oils are difficult to wash from clothes because they do not dissolve in water.
- Fats and oils are used to make soap when they are split up by hot sodium hydroxide:
  fat + sodium hydroxide ⟶ soap + glycerol
- In soap manufacture, vegetable oils are heated in large vats with sodium hydroxide solution. This stage is called **saponification**.
- Salt is added at the end of the reaction to make the soap precipitate out.
- The solid soap can then be removed and colouring and perfume may then be added.

water–in–oil emulsion contains droplets of water spread through oil

oil–on–water emulsion contains droplets of oil spread through water

Saponification

- The esters that make up fats and oils have the structure shown here.
- When these react with sodium hydroxide, the molecules split into a molecule of glycerol and molecules of soap. fat + sodium hydroxide ⟶ soap + glycerol
- This reaction is the reaction of water in alkaline conditions, so it is a hydrolysis reaction.
- As this particular hydrolysis reaction makes soap, it is called a saponification reaction.

Questions

Grades D–C
1 How is bromine used to test for unsaturation?

Grades B–A*
2 Why are unsaturated fats thought to be a healthier option in a diet.

Grades D–C
3 What happens during saponification?

Grades B–A*
4 Explain how a saponification reaction works.

Analgesics

Medicines and side effects

Grades D–C

- The active ingredients of medicines are drugs. A drug is any externally administered chemical which affects the body's chemical reactions. We say 'externally administered' because the body also produces chemicals of its own which affect its reactions.
- The chemicals used to make the analgesics are very pure.
- An overdose of aspirin causes severe bleeding in the stomach.
- An overdose of paracetamol causes liver damage.
- If several drugs are being taken, the drugs may interfere with each other. It is always important to warn your doctor if you are already taking any other medicine.

Grades B–A*

- People will always react with slightly different side effects to powerful drugs.
- Aspirin can attack the stomach lining and cause slight bleeding in the stomach.
- Some people are allergic to aspirin and can develop severe asthma attacks with it.
- Aspirin is no longer used in most children's medicines. This does not mean that aspirin is more dangerous than other drugs. All drugs have different side effects.

Formulae

Grades D–C

- The formula of drugs.
- The molecular formula of paracetamol is $C_8H_9NO_2$.
- Aspirin is now sold in the form of soluble aspirin.
- Soluble aspirin dissolves in a glass of water, which makes it easier to swallow and faster to act. It also has fewer side effects than aspirin itself.

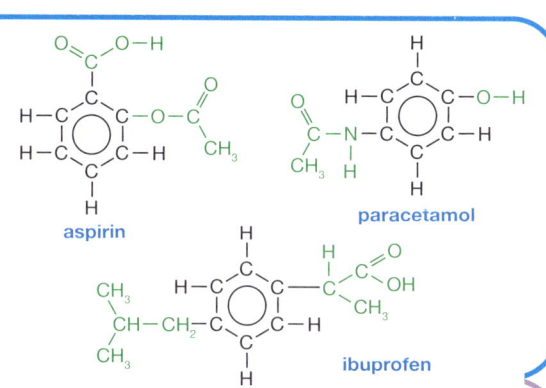

The formulae of three common analgesics

Grades B–A*

- All three analgesics contain a carbon ring structure known as a benzene ring. Also, they have an alcohol group –OH, an acid group, –COOH, and a CH_3CO– group.
- Covalently bonded molecules do not dissolve in water, so aspirin is not very soluble.
- Aspirin contains the acid group –COOH. Soluble aspirin is made by reacting aspirin with a base such as sodium hydroxide or calcium carbonate.
- The –COOH group forms the –COO– ion, this ionic part makes aspirin soluble in water.
- Salicylic acid and ethanoic anhydride are heated together in batches. The product is then slowly cooled to allow the aspirin, which is not very soluble, to form large crystals.
- The crystals are then filtered off before going to a purification stage.

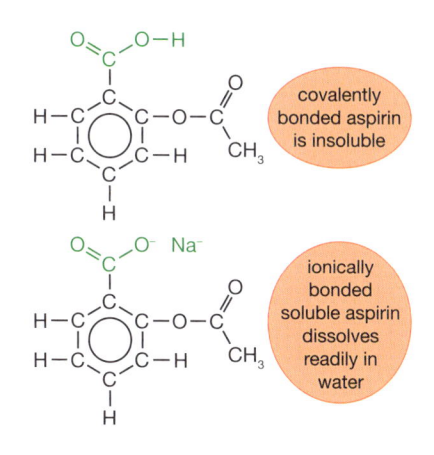

covalently bonded aspirin is insoluble

ionically bonded soluble aspirin dissolves readily in water

Questions

Grades D–C

1. What is the danger of an overdose of paracetamol?

Grades B–A*

2. What is an unwanted side effect of aspirin?

Grades D–C

1. What is the molecular formula of ibruprofen?

Grades B–A*

4. Explain the chemical reaction that makes aspirin soluble?

C6 Summary

C6 CHEMISTRY OUT THERE

Fuel cells

The reaction between hydrogen and oxygen is an **exothermic reaction**.

The electrode reactions in the fuel cell are:
$2H_2 \longrightarrow 4H^+ + 4e^-$
$2H_2O + O_2 + 4e^- \longrightarrow 4OH^-$

A **fuel cell** is supplied with hydrogen and oxygen and uses the energy from the reaction between them to create electrical energy.

The advantages of a hydrogen–oxygen fuel cell for generating electricity are:
– greater efficiency
– fewer stages
– less pollution
– direct energy transfer.

The reaction in a hydrogen–oxygen fuel cell is: $2H_2 + O_2 \longrightarrow 2H_2O$

Alcohols

The conditions for fermentation are needed because:
– if the temperature is too low the yeast is inactive
– if it is too high the yeast enzymes are denatured
– the absence of air prevents ethanoic acid forming.

Pentanol has five carbon atoms. Its molecular formula is $C_5H_{11}OH$.

Glucose $\longrightarrow$ carbon dioxide + ethanol
This is **fermentation**.
Alcohol can also be made from ethene.
ethene + water $\longrightarrow$ ethanol

The **molecular formula** of ethanol is C_2H_6O. The **displayed formula** of ethanol is:

H H
| |
H—C—C—O—H
| |
H H

Ethanol can be dehydrated to make ethene by passing ethanol vapour over a heated catalyst. The catalyst is aluminium oxide.
ethanol $\longrightarrow$ ethene + water

The ozone layer

CFCs have been banned in the UK as they lead to the depletion of the **ozone** layer.

Ultraviolet light on CFCs make **chlorine** atoms. The making of the chlorine atoms in the **stratosphere** leads to the depletion of the ozone layer.

More **ultraviolet** light can reach the surface of the Earth if the ozone layer is depleted.

Only a small number of chlorine atoms are needed to cause a great deal of damage as a chain reaction is set up.

Analgesics

A **drug** is an externally administered substance which modifies or affects chemical reations in the body.

Soluble aspirin is already dissolved so it is absorbed faster and has less time to do damage to the lining of the stomach. Soluble aspirin is made by reacting aspirin with a base such as sodium hydroxide. The –COOH group forms the –COO⁻ ion, making the molecule ionic and so soluble in water.

This is the displayed formula of aspirin.

This is the displayed formula of ibruprofen. Notice that it also has a benzene ring and a –COOH group.

How science works

Understanding the scientific process

As part of your Chemistry assessment, you will need to show that you have an understanding of the scientific process – How Science Works.

This involves examining how scientific data is collected and analysed. You will need to evaluate the data by providing evidence to test ideas and develop theories. Some explanations are developed using scientific theories, models and ideas. You should be aware that there are some questions science cannot answer.

Collecting and evaluating data

You should be able to devise a plan that will answer a scientific question or solve a scientific problem. In doing so, you will need to collect data from both primary and secondary sources. Primary data will come from your own findings – often from an experimental investigation. Whilst working with primary data, you will need to show that you can work safely and accurately, not only on your own but also with others.

Secondary data is found by research – often using ICT but do not forget books, journals, magazines and newspapers are also sources of secondary data. The data you collect will need to be evaluated for its validity and reliability.

Presenting information

You should be able to present your information in an appropriate, scientific manner. This means being able to develop an argument and come to a conclusion based on recall and analysis of scientific information. It is important to use both quantitative and qualitative arguments.

Changing ideas and explanations

Many of today's scientific and technological developments have both benefits and risks. The decisions that scientists make will almost certainly raise ethical, environmental, social or economic questions. Scientific ideas and explanations change as time passes and it is the job of scientists to validate these changing ideas.

In 1692, the British astronomer Edmund Halley (after whom Halley's Comet was named) suggested that the Earth consisted of four concentric spheres. He was trying to explain the magnetic field that surrounds the Earth. There was, he said, a shell about 500 miles thick, two inner concentric shells and an inner core. The shells were separated by atmospheres and each shell had magnetic poles. The spheres rotated at different speeds. He believed this explained why unusual compass readings occurred. He also believed that each of these inner spheres supported life which was constantly lit by a luminous atmosphere.

This may sound quite an absurd idea today, but it is the work of scientists for the past 300 years that has developed different models that are constantly being refined.

How science works

Science in the News

Assessment

Science in the News is intended as the main way in which the OCR Chemistry course assesses your understanding of How Science Works.

Whilst some of you will continue to study science, many of you will have completed your science education by the time you have finished your GCSE course. It is important that you are able to meet any scientific challenge which arises in later life.

It is important that you realise when data or information is not presented in an accurate way. Think about what is wrong in this example based on a newspaper article.

Acid Alert Closes Motorway
by John Smith

The M6 motorway near Stafford was closed earlier today after a tanker carrying 20 tonnes of highly acidic sodium hydroxide solution started to leak some of the contents.

Awareness of current issues

You should also be aware of what aspects of science may be important for people living in the 21st Century.

One of the most controversial topics at this time is the most effective use of the dwindling supplies of crude oil.

Your Science in the News assessment will ask you to undertake some research on a scientific issue. The task set to you will be in the form of a question. You will then have to produce a short report which will clearly show that you have
- considered both sides of the argument
- decided on the suitability, accuracy and/or reliability of the evidence
- considered the impact on society and the environment
- justified your conclusion about the question asked.

The aim is to equip you with life-long skills that will allow you to take a full and active part in the scientific 21st Century.

Collins Revision

GCSE Higher Chemistry

Exam Practice Workbook

FOR OCR GATEWAY B

Fundamental concepts

1 a Construct a **balanced symbol** equation for the reaction between magnesium (Mg) and oxygen (O_2) to make magnesium oxide (MgO).
_____ [2 marks]

b Construct a **balanced symbol** equation for the reaction between copper oxide (CuO) and nitric acid (HNO_3) to make copper nitrate ($Cu(NO_3)_2$) and water (H_2O).
_____ [2 marks]

c Construct a **balanced symbol** equation for the reaction between potassium chloride and lead(II) nitrate to make potassium nitrate and lead chloride.
_____ [2 marks]

2 a i Write down the total number of atoms in the formula of copper nitrate:
($Cu(NO_3)_2$). _____ [1 mark]

ii Write down the number of each different type of atom in $Cu(NO_3)_2$.
_____ and _____ and _____ [1 mark]

b Write down the formula of these substances.

i sulfuric acid _____ hydrochloric acid _____
nitric acid _____ ethanoic acid _____ [4 marks]

ii ammonia _____ sodium hydroxide _____
potassium hydroxide _____ [3 marks]

iii sodium chloride _____ magnesium sulfate _____
calcium chloride _____ potassium sulfate _____ [4 marks]

iv barium chloride _____ carbon dioxide _____
hydrogen _____ water _____ [4 marks]

v sodium carbonate _____ ethanol _____
glucose _____ [3 marks]

3 a Look at the displayed formula of ethanoic acid.

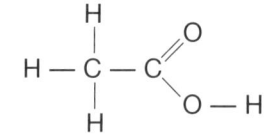

i What do the lines joining the atoms represent?
_____ [1 mark]

ii Write down the molecular formula of ethanoic acid. _____ [1 mark]

b Atoms in this acid are held together by covalent bonds. What is the other **type** of bond? _____ [1 mark]

c Describe a covalent bond. Use ideas about electrons.
_____ [1 mark]

d Construct a balanced equation using these **displayed** formulae.

$H-\underset{H}{\overset{H}{C}}-H + O=O \rightarrow O=C=O + HH$ (with O above)

[1 mark]

Cooking

1 Some foods need to be cooked. Explain why.

_____

_____ [2 marks]

2 a Write down **two** good sources of carbohydrate.

_____ [2 marks]

b Write down **two** good sources of protein.

_____ [2 marks]

c Protein molecules change when they are cooked. Explain why this is important.

_____

_____

_____ [3 marks]

3 a Baking powder is a chemical called **sodium hydrogencarbonate**. When it is heated it **decomposes** to give sodium carbonate, carbon dioxide and water.
  i Write down the word equation for the reaction.

_____

_____ [2 marks]

  ii Write down the **reactants** of the reaction.

_____ [1 mark]

  iii Write down the **products** of the reaction.

_____ [1 mark]

b The formula for sodium hydrogencarbonate is $NaHCO_3$. Write down the balanced symbol equation for this reaction.

_____

_____ [3 marks]

4 The chemical test for carbon dioxide is to pass it through limewater. It will turn the limewater from _____ to _____ .

[2 marks]

C1 CARBON CHEMISTRY

Food additives

1 a Why are antioxidants added to tinned fruit?
_____ [1 mark]

b Ascorbic acid (vitamin C) is used as an antioxidant in which foods?
_____
_____ [2 marks]

c Information about food is given on food labels. How much energy does this food provide?

Energy	56 J
Sodium	0.02 mg
Magnesium	Trace

_____ [1 mark]

2 a Why is food packaging used?
_____ [1 mark]

b What is active packaging?
_____ [1 mark]

c Why is active packaging beginning to be used?
_____ [1 mark]

d It uses a **polymer**, what else is needed to make it work?
_____ [1 mark]

e Intelligent packaging works by including **indicators** on packages. Explain how these work.
_____ [1 mark]

3 a Look at the diagram.
It is a detergent molecules made up of two parts, a head and a tail. Describe how the detergent works on removing grease in water. Use the diagram to help you.

_____
_____
_____
_____ [5 marks]

b Mayonnaise is made of oil, vinegar and egg yolk. This makes a smooth substance. Explain how. Use the words **hydrophilic** and **hydrophobic** in your answer. Use a diagram to help you.

_____
_____
_____
_____ [5 marks]

Smells

1 a To make a perfume, alcohol is mixed with an acid to make an ester.

 b Write down a word equation for this reaction.
_____ [2 marks]

 c Look at the diagram. Label the alcohol and acid. [1 mark]

 d Label the condenser. [1 mark]

 e What is happening at **X**?
_____
_____ [1 mark]

 f Why is the condenser used?
_____
_____
_____ [1 mark]

2 A good perfume needs to have several properties. These are listed in the boxes. Draw a **straight** line to match the **best** reason to the property needed.

evaporates easily	it can be put directly on the skin
non-toxic	its particles can reach the nose
insoluble in water	it does not poison people
does not irritate the skin	it cannot be washed off easily

[3 marks]

3 a A solute and a solvent that do not separate is a _____ . [1 mark]

 b Esters are used as perfumes and _____ . [1 mark]

4 a If a liquid evaporates easily then the substance is volatile. Explain how this happens. Use ideas about forces between particles in your answer.
_____
_____
_____
_____ [4 marks]

 b Water does not dissolve nail varnish. Use the diagram to help explain why.
Use ideas about the forces of attraction between molecules in your answer.
_____
_____
_____
_____ [2 marks]

Making crude oil useful

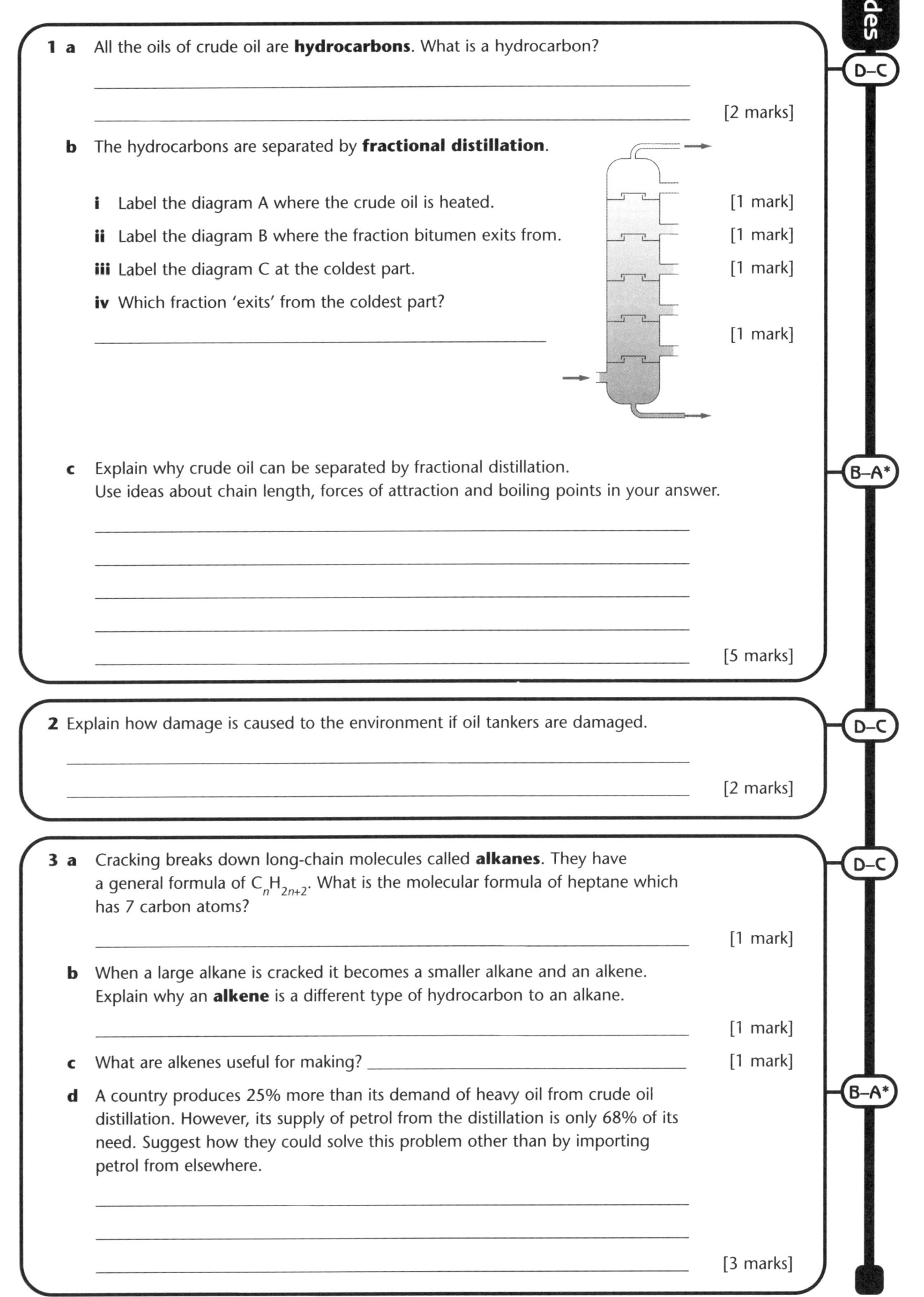

1 a All the oils of crude oil are **hydrocarbons**. What is a hydrocarbon?

_____

_____ [2 marks]

b The hydrocarbons are separated by **fractional distillation**.

   i Label the diagram A where the crude oil is heated. [1 mark]

   ii Label the diagram B where the fraction bitumen exits from. [1 mark]

   iii Label the diagram C at the coldest part. [1 mark]

   iv Which fraction 'exits' from the coldest part?

_____ [1 mark]

c Explain why crude oil can be separated by fractional distillation.
Use ideas about chain length, forces of attraction and boiling points in your answer.

_____

_____

_____

_____

_____ [5 marks]

2 Explain how damage is caused to the environment if oil tankers are damaged.

_____

_____ [2 marks]

3 a Cracking breaks down long-chain molecules called **alkanes**. They have a general formula of C_nH_{2n+2}. What is the molecular formula of heptane which has 7 carbon atoms?

_____ [1 mark]

b When a large alkane is cracked it becomes a smaller alkane and an alkene. Explain why an **alkene** is a different type of hydrocarbon to an alkane.

_____ [1 mark]

c What are alkenes useful for making? _____ [1 mark]

d A country produces 25% more than its demand of heavy oil from crude oil distillation. However, its supply of petrol from the distillation is only 68% of its need. Suggest how they could solve this problem other than by importing petrol from elsewhere.

_____

_____

_____ [3 marks]

Making polymers

1 a Which molecule is a polymer? Put a ring around **A**, **B**, **C** or **D**.

A: H-C(H)(H)-C(H)(H)-OH

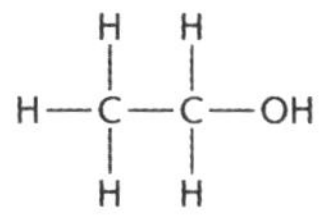

B: H-C(H)(H)-C(H)(H)-C(H)(H)-H

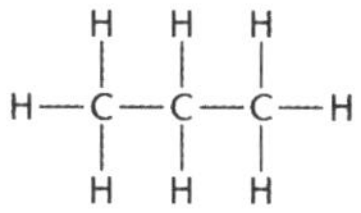

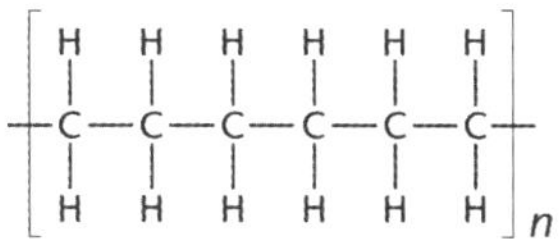

C: [-C(H)(H)-C(H)(H)-C(H)(H)-C(H)(H)-C(H)(H)-C(H)(H)-]$_n$

D: C(H)-C(H)=C-Cl

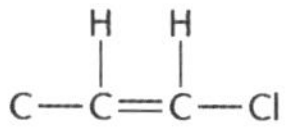

[1 mark]

b Write down **two** conditions needed for polymerisation.

_____
_____ [2 marks]

c What does a monomer need in its structure to undergo addition polymerisation?

_____ [1 mark]

d Construct the **displayed formula** of the addition polymer from this monomer. [2 marks]

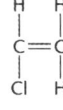

C(H)(Cl)=C(H)(H)

e Draw the monomer that makes this addition polymer. [2 marks]

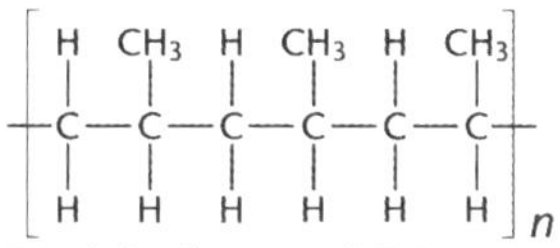

[-C(H)(H)-C(CH$_3$)(H)-C(H)(H)-C(CH$_3$)(H)-C(H)(H)-C(CH$_3$)(H)-]$_n$

f Explain how addition polymerisation takes place.

_____
_____
_____
_____ [4 marks]

2

a i Butanol, C$_4$H$_9$OH, H-C(H)(H)-C(H)(H)-C(H)(H)-C(H)(H)-OH is **not** a hydrocarbon. Explain why.

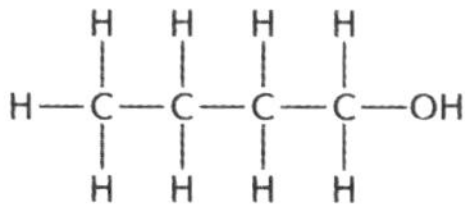

_____ [1 mark]

ii Butene is an alkene.

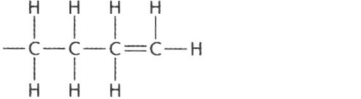

H-C(H)(H)-C(H)(H)-C(H)=C(H)-H

Explain how you know.

_____ [1 mark]

iii Butene is a **monomer**. What is **polybutene**?

_____ [1 mark]

b i Bromine solution is used to test for unsaturation. It is an orange solution. When an alkene is added the orange solution turns colourless. Explain why.

_____ [1 mark]

ii When bromine solution is added to an alkane, what do you see?

_____ [1 mark]

Designer polymers

1 a Polymers are better than other materials for some uses. Give **one** example of the use of a polymer in healthcare that is better than the material used before.

_____

_____ [2 marks]

b i Gore-Tex® type materials are used to make clothing waterproof and breathable. The inner layer of the clothing is made from expanded PTFE (polytetrafluoroethene) which is **hydrophobic**. Explain what hydrophobic means.

_____

_____ [2 marks]

ii The PTFE is expanded to form a **microporous membrane**. Explain how this makes a breathable membrane.

_____

_____ [2 marks]

2 a Scientists are developing addition polymers that are **biodegradable**. Explain why.

_____

_____ [2 marks]

b Suggest a use for a biodegradable plastic.

_____ [1 mark]

c Disposing of non-biodegradable polymers causes problems. Explain the problems for each of the ways of disposing.

Landfill sites _____

Burning waste plastic _____

Recycling _____ [3 marks]

3 Look at the diagram of polymer chains.

a i Label the **intramolecular covalent** bonds. [1 mark]

ii Label the **intermolecular** forces of attraction. [1 mark]

b Some plastics have low melting points and can be stretched easily whereas other plastics have high melting points and cannot be stretched easily. They are rigid. Explain why. Use ideas about forces between chains in your answer. You may use a diagram to help you.

_____

_____

_____

_____ [4 marks]

Using carbon fuels

1 a i Look at the table. Which fuel produces more acid fumes?

characteristic	coal	petrol
energy value	high	high
availability	good	good
storage	bulky and dirty	volatile
toxicity	produces acid fumes	produces less acid fumes
pollution caused	acid rain, carbon dioxide and soot	carbon dioxide, nitrous oxides
ease of use	easy in power stations	easy in engines

_____ [1 mark]

ii Give **two** advantages of using either coal or petrol for heating.

_____
_____ [2 marks]

b Explain why the use of petrol and diesel in transport is contributing to a global problem.

_____
_____
_____
_____ [4 marks]

2 a Write down a **word equation** for a hydrocarbon fuel burning in air.

_____ [1 mark]

b Two products are made in the complete combustion of a fuel. These can be tested in the laboratory.
  i Limewater is used to test for one product. Which one?

_____ [1 mark]

  ii How is the other product tested?

_____ [1 mark]

c Complete combustion is better than incomplete combustion. Explain why.

_____
_____
_____ [3 marks]

d Why should a room be well ventilated and a heater regularly checked?

_____
_____ [2 marks]

e Write a balanced equation for the complete combustion of propane, C_3H_8.

_____ [2 marks]

Energy

1 a Use the words **exothermic** and **endothermic** correctly in these sentences.

When energy is transferred **out** to the surroundings in a chemical reaction it is an

_____ reaction (energy is released).

When energy is taken in from the surroundings in a chemical reaction it is an

_____ reaction (absorbs energy).

An_____ reaction is shown by a temperature **increase**.

Burning magnesium is an example of an _____ reaction. [4 marks]

b Is bond breaking an exothermic or endothermic process?

_____ [1 mark]

c Burning methane is an exothermic reaction. Explain why, using ideas about bond breaking and bond making.

_____

_____

_____ [3 marks]

2 The flame of a Bunsen burner changes colour depending on the amount of oxygen that it burns in. The flame is _____ when the gas burns in plenty of oxygen. This is because there is _____ combustion. The flame is _____ when the gas burns in limited oxygen. This is because there is _____ combustion. [4 marks]

3 a Describe how you would design your own experiment to compare the energy transferred by two different fuels, using a beaker of water and a thermometer. Write down the ways to make the tests fair.

_____

_____

_____

_____ [5 marks]

b Todd and Terri calculate the amount of energy transferred during a reaction. They use a spirit burner to burn the fuel to heat 100 g of water in a copper **calorimeter**. They measure a temperature change in the water of 50 °C. The mass of fuel they burn is 4.00 g.

   i How do they make the tests reliable?

   _____ [1 mark]

   ii Show how they calculate the energy released per gram.
   (The specific heat capacity of water is 4.2 J/g/°C)

[5 marks]

70

C1 Revision checklist

- I know that cooking food is a chemical change as a new substance is made and it is an irreversible reaction. ☐

- I know that protein molecules in eggs and meat change shape when the food is cooked. ☐

- I know that when the shape of a protein changes it is called denaturing. ☐

- I know that emulsifiers are molecules that have a water-loving part and an oil- or fat-loving part. ☐

- I know that alcohols react with acids to make an ester and water. ☐

- I know that a solute is the substance dissolved in a solvent to make a solution that does not separate. ☐

- I know that crude oil is a non-renewable fossil fuel, which is a mixture of many hydrocarbons. ☐

- I know that petrol is a crude oil fraction with a low boiling point, which exits at the top of the fractional distillation tower. ☐

- I know that polymerisation is a process which requires high pressure and a catalyst. ☐

- I know that a hydrocarbon is a compound formed between carbon atoms and hydrogen atoms only. ☐

- I know that alkenes are hydrocarbons with one or more double bonds between carbon atoms. ☐

- I know that complete combustion of a hydrocarbon fuel makes carbon dioxide and water only. ☐

- I know that an exothermic reaction is one where energy is released into the surroundings. ☐

- I can work out that: energy transferred = mass of water × 4.2 × temperature change. ☐

C1 CARBON CHEMISTRY

Paints and pigments

1 a An **emulsion paint** is a water-based paint. Explain how it covers a surface.

_____

_____ [2 marks]

b Sam and Chris are discussing why paint is a colloid. Sam gives an explanation. He uses ideas about particle size and mixtures and dispersion. Write down what he says.

_____

_____

_____

_____ [4 marks]

c The oil in oil paint is very sticky and takes a long time to harden. Explain how it hardens in two stages and what it forms.

_____

_____

_____ [3 marks]

2 a Thermochromic pigment changes colour at 45 °C. Write down **two** examples it is used for.

_____

_____ [2 marks]

b Most thermochromic pigments change from a particular colour to colourless. How do they ensure a greater range of colours is available?

_____ [1 mark]

c Thermochromic paints come in a limited range of colours. If a green paint becomes yellow when heated explain what the mixture contains and how it changes colour.

_____

_____

_____

_____ [4 marks]

d A pigment that stores absorbed energy is called

_____ [1 mark]

e What does it release that energy as?

_____ [1 mark]

3 a Where are phosphorescent pigments used?

_____ [1 mark]

b What have phosphorescent pigments replaced and why?

_____

_____ [2 marks]

Construction materials

1 a Put these materials into the order of hardness.

        granite   limestone   marble

Least hard _____ _____ _____ Hardest   [3 marks]

b Brick, concrete, steel, aluminium and glass are manufactured.
Finish the table to show the raw materials they come from.

building material	brick	cement	glass	iron	aluminium
raw material					

[5 marks]

c **Igneous**, **sedimentary** and **metamorphic** rocks differ in hardness.

Which rock is usually the least hard? _____ [1 mark]

d Granite and marble are different types of rock. Their hardness is different. Compare how granite and marble are made and why their structures are different. Use **all** the words in this list.

crystals igneous interlocking heat pressure slowly metamorphic solidifies

granite	marble

[8 marks]

e Limestone is a **sedimentary rock**. Explain how limestone is made.

_____
_____ [3 marks]

2 a Calcium carbonate decomposes at a very high temperature. Write a word equation.

_____ [3 marks]

b Cement is made from limestone. Write down how.

_____ [2 marks]

c Write down the symbol equation for the thermal decomposition of calcium carbonate.

_____ [3 marks]

d Reinforced concrete is a better construction material than non-reinforced concrete. Explain why. Use ideas about stretching and tension in concrete and strength under tension of steel.

_____
_____
_____
_____ [4 marks]

Does the Earth move?

1 a Are the tectonic plates that make up the Earth's crust less dense or more dense than the mantle?

_____ [1 mark]

b Write down the **two** kinds of tectonic plate.

_____ and _____ [2 marks]

c The mantle is always solid, but at greater depths it is more like Plasticine, which can 'flow'. The **tectonic plates** move very slowly on this mantle. They can move in different ways. Explain how, using diagrams to help you explain.

[5 marks]

d Explain how subduction occurs.

_____
_____
_____
_____
_____ [6 marks]

e Write down **two** pieces of evidence used to develop the theory of plate tectonics. Use ideas about continental drift and mid-ocean ridges.

_____
_____
_____ [4 marks]

2 a Magma can rise through the Earth's crust. Explain why. [1 mark]

_____

b Magma cools and solidifies into igneous rock either after it comes out of a volcano as lava, or before it even gets to the surface. By looking at crystals of igneous rock, geologists can tell how quickly the rock cooled. Fill in the **two** boxes with an explanation and an example.

basalt cools rapidly	cools slowly granite
small crystals	large crystals

[4 marks]

c i There are two types of magma, iron-rich and silica-rich. Which is the magma that is less runny and produces volcanoes that may erupt explosively?

_____ [1 mark]

ii Explain what it produces.

_____
_____ [2 marks]

C2 ROCKS AND METALS

Metals and alloys

1 a Copper used for recycling has to be sorted carefully so that valuable 'pure' copper scrap is not mixed with less pure scrap. When impure copper is used to make alloys what must happen first?

_____ [1 mark]

b If the scrap copper is very impure what must be done before it is used again?

_____ [1 mark]

2 Impure copper can be purified in the laboratory using an electrolysis cell.

a What is the **anode** made from?

_____ [1 mark]

b What happens at the **cathode**?

_____ [1 mark]

c What happens to the **anode**?

_____ [1 mark]

d What happens to the thickness of the cathode?

_____ [1 mark]

e What is the impure copper called?

_____ [1 mark]

f What happens to the impurities from the copper anode?

_____ [1 mark]

3 a Most metals form alloys. Draw a **straight** line to match the metals to the alloy.

amalgam	contains copper and zinc
solder	contains mercury
brass	contains lead and tin

[2 marks]

b Alloys are often more useful than the original metals, though nowadays pure copper is more important than bronze or brass. Why are vast amounts turned into electric wire?

_____ [1 mark]

c Write down **two** properties of smart alloys.

_____
_____ [2 marks]

d Write down **two** new ways of using smart alloys.

_____
_____ [2 marks]

e **Nitinol** is a smart alloy. Which **two** metals is it made from?

_____
_____ [2 marks]

Cars for scrap

1 a In winter, icy roads are treated with salt. Why is this a problem for steel car bodies?
_____ [1 mark]

b Aluminium does not corrode in moist air. Explain why.
_____ [1 mark]

c Rust is an oxide layer but it does not protect the rest of the iron. Explain why.
_____ [1 mark]

d Rusting is a **chemical reaction** between iron and oxygen to make an oxide. The chemical name for rust is **hydrated iron(III) oxide**.
Write down the word equation for rusting.

_____ [3 marks]

2 a Steel is an alloy made of _____ and _____. [2 marks]

b Write down **two** advantages of steel over iron.
_____ [2 marks]

c Steel and aluminium can both be used to make car bodies but each material has its own advantages. Write down **two** advantages of cars made from aluminium and **one** disadvantage.

_____
_____
_____ [3 marks]

d Finish this table. Write down the reasons why these materials are used.

material and its use	reasons material is used
aluminium in car bodies and wheel hubs	
copper in electrical wires	
plastic in dashboards, dials, bumpers	
pvc in metal wire coverings	
plastic/glass composite in windscreens	
fibre in seats	

[6 marks]

3 Write down **three** benefits that recycling metals and the other materials of a car have on the environment.

_____
_____ [3 marks]

Clean air

1 a Label the pie chart with the four main gases of the air in the correct section. [4 marks]

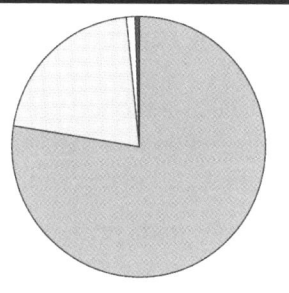

b Mark in the percentages of the gases. [4 marks]

c These percentages do not change very much because there is a balance between three of the processes that use up or make carbon dioxide and use or make oxygen. Explain how the balance is maintained.

_____
_____
_____
_____ [7 marks]

2 a Over the last few centuries the percentage of carbon dioxide in air has increased slightly. This is due to a number of factors. Explain these factors.

increased energy usage	
increased population	
deforestation	

[3 marks]

b Explain **one** theory of how the Earth's atmosphere has evolved. Include in your answer
  i the original gases of the atmosphere
  ii how nitrogen levels increased
  iii how the levels of carbon dioxide and oxygen evolved to those which they are today.

_____
_____
_____ [4 marks]

3 a Finish the table to describe the origin of these atmospheric pollutants.

pollutant	carbon monoxide	oxides of nitrogen	sulfur dioxide
origin of pollutant			

[3 marks]

b Which metal catalyst do catalytic converters contain?

_____ [1 mark]

c i A reaction between nitric oxide and carbon monoxide takes place on the surface of the catalyst. The reaction forms nitrogen and carbon dioxide. Why is it important that these two gases are made?

_____ [1 mark]

ii Write down the word equation for this reaction.

_____ [2 marks]

iii Write down the balanced symbol equation for this reaction.

_____ [3 marks]

Faster or slower (1)

1 a Look at **Graph A**. It shows the reaction between magnesium and acid at 20 °C.

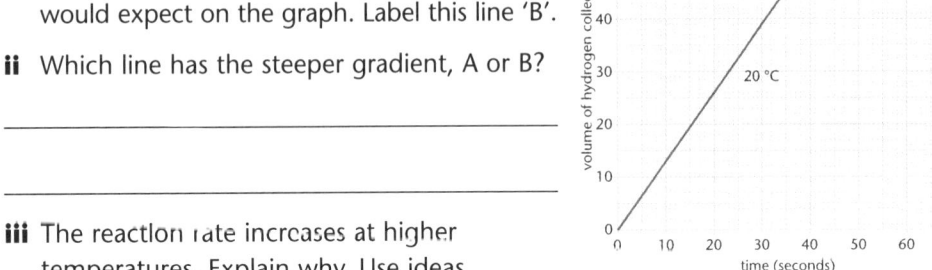

  i If the reaction takes place at a higher temperature mark the reaction line that you would expect on the graph. Label this line 'B'. [2 marks]

  ii Which line has the steeper gradient, A or B?

  _____

  _____ [1 mark]

  iii The reaction rate increases at higher temperatures. Explain why. Use ideas about particles in your answer.

  _____

  _____ [4 marks]

b The rate of a reaction is not determined by the number of collisions. Explain what it is determined by.

  _____

  _____ [2 marks]

c Explain the increase in reaction rate in terms of particle collisions.

  _____ [1 mark]

2 Look at **Graph B**. It shows the reaction between magnesium and acid of different concentrations.

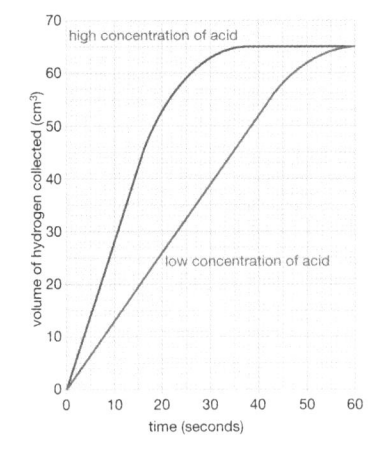

a What does the graph show about

  i the reaction rates with the two acids?

  _____

  _____

  ii the amount of hydrogen collected in the two reactions?

  _____

  _____ [4 marks]

b The rate of reaction can be worked out from the gradient of **Graph C**.

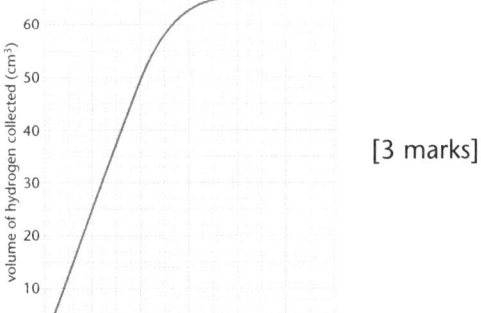

  i Draw construction lines on the graph and calculate the gradient of the line.

  _____ [3 marks]

  ii What does **interpolation** mean?

  _____

  _____ [1 mark]

Faster or slower (2)

1 What products are made during an explosion?

_____ [2 marks]

2 a The reaction between calcium carbonate and hydrochloric acid is measured by the decrease in mass. Look at the equation, why is there a decrease in mass?

$CaCO_3 + 2HCl \rightarrow CaCl_2 + H_2O + CO_2$

_____ [2 marks]

b The graph shows how the rate of reaction between calcium carbonate and dilute hydrochloric acid is measured.

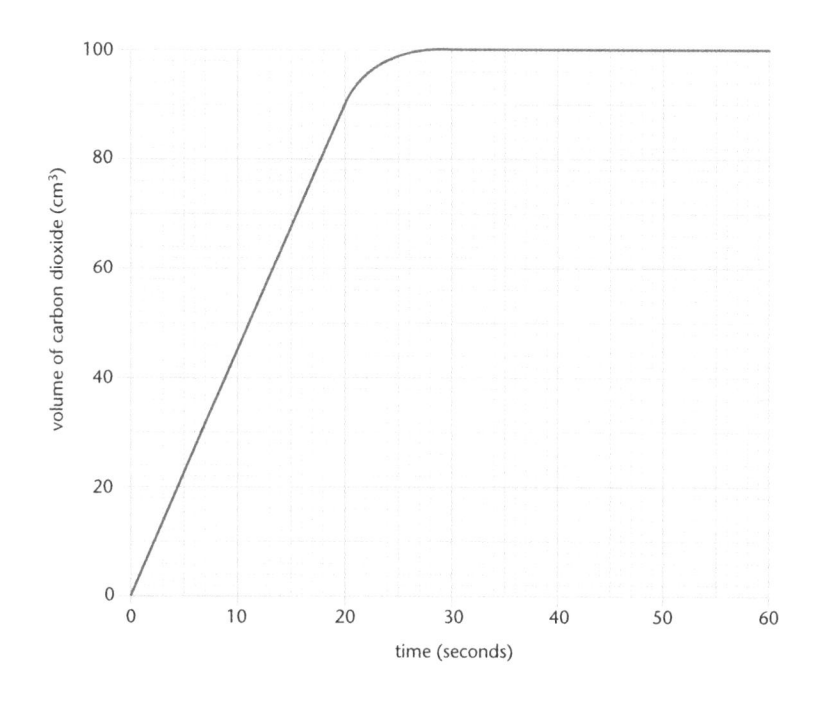

[Graph: volume of carbon dioxide (cm³) vs time (seconds). Curve rises steeply from 0 to about 95 cm³ by 25 seconds, then levels off at 100 cm³.]

i At which time does the reaction stop? _____ [1 mark]

ii If this reaction were using powdered calcium carbonate, sketch on the graph the line that would show the reaction of the same mass of calcium carbonate as lumps. [2 marks]

iii Why does the reaction slow down? Use ideas about collisions in your answer.

_____ [1 mark]

c Why does the reaction rate increase if the surface area increases. Use ideas about collisions between particles in your answer.

_____ [1 mark]

3 a What does a catalyst do?

_____ [1 mark]

b Most catalysts only make a **specific** reaction faster. Explain why.

_____
_____ [2 marks]

C2 Revision checklist

C2 ROCKS AND METALS

- I know that paint is a colloid where solid particles are dispersed in a liquid, but are not dissolved. ☐

- I know that thermochromic pigments change colour when heated or cooled. ☐

- I know that brick is made from clay, glass from sand, and aluminium and iron from ores. ☐

- I know that the equation for the decomposition of limestone is:
  calcium carbonate → calcium oxide + carbon dioxide. ☐

- I know that the outer layer of the Earth is made up of continental plates with oceanic plates under oceans. ☐

- I know that the size of crystals in an igneous rock is related to the rate of cooling of molten rock. ☐

- I know that copper can be extracted by heating its ore with carbon, but purified by electrolysis. ☐

- I know that alloys are mixtures of metals, for example, copper and zinc make brass. ☐

- I know that aluminium does not corrode when wet as it has a protective layer of aluminium oxide. ☐

- I know that iron rusts in air and water to make hydrated iron(III) oxide. ☐

- I know that toxic carbon monoxide comes from incomplete combustion of petrol or diesel in cars. ☐

- I know that catalytic converters can remove CO by conversion:
  $2CO + 2NO \rightarrow N_2 + CO_2$. ☐

- I know that a temperature increase makes particles move faster, so increasing the rate of reaction. ☐

- I know that a catalyst is a substance which changes the rate of reaction but is unchanged at the end. ☐

What are atoms like?

1 a What are the particles in the nucleus of an atom?
_____ [1 mark]

b Finish the table to show the relative mass and charge of the atomic particles.

	relative charge	relative mass
electron		0.0005 (zero)
proton	+1	
neutron		

[4 marks]

c What is the atomic number of an atom?
_____ [1 mark]

d What is the mass number?
_____ [1 mark]

e What is the atom that has an atomic number of 9, and a mass number of 19, and a neutral charge?

atomic number	9
mass number	19
charge	0
name	

f If an element has the symbol $^{35}_{17}Cl$ how many protons are there in the atom? How many neutrons?

  i Protons _____ [1 mark]

  ii Neutrons _____ [1 mark]

2 a What is an isotope?
_____ [1 mark]

b Isotopes of an element have different numbers of neutrons in their atoms. Finish the table.

isotope	electrons	protons	neutrons
$^{12}_{6}C$	6	6	
$^{14}_{6}C$	6	6	

[2 marks]

3 Draw the electronic structure for the element aluminium. Explain why **three** shells are needed for electrons.

_____
_____
_____
_____
_____ [5 marks]

Ionic bonding

1 a Put a tick (✓) in the box next to the sentence that describes a **metal** atom.

An **atom** that has extra electrons in its outer shell and needs to **lose** them to be stable. ☐

An **atom** that has 'spaces' in its outer shell and needs to **gain** electrons to be stable. ☐

[1 mark]

b Draw a diagram to show how the electrons transfer between a metal atom and a non-metal atom to form a stable pair. **Outer shells only**.

[3 marks]

c Finish the sentences.

  i If an atom loses electrons a _____ **ion** is formed. [1 mark]

  ii An example of an atom which loses 2 electrons is _____. [1 mark]

d Finish the sentences.

  i A **negative ion** is formed by an atom _____ electrons. [1 mark]

  ii An example of an atom gaining 1 electron is _____. [1 mark]

e Finish the sentences.
During **ionic bonding**, the metal atom becomes a _____ ion and the non-metal atom becomes a _____ ion.
The positive ion and the negative ion then attract one another. They attract to a number of other ions to make a solid _____. [3 marks]

f Draw the **'dot and cross' model** for sodium chloride. **Outer shell electrons only.**

[4 marks]

g Draw the dot and cross diagram for magnesium chloride.

[4 marks]

h Put a tick (✓) in the boxes next to the substances that conduct electricity.

sodium chloride solution ☐   solid sodium chloride ☐

molten (melted) magnesium oxide ☐   solid magnesium oxide ☐

molten sodium chloride. ☐

[3 marks]

i Finish the sentences for some of the physical properties of sodium chloride and magnesium oxide.

  i There is a strong attraction between positive and negative ions so they have _____ melting points.

  ii Its ions cannot move in the solid so it does not _____.

  iii Solutions or molten liquids conduct electricity because _____. [3 marks]

Covalent bonding

1 a Non-metals combine together by **sharing** electrons. What is this type of bonding?

_____ [1 mark]

b Look at the diagram.

Explain how a water molecule is formed from atoms of other elements.

_____
_____
_____ [4 marks]

c Carbon dioxide and water do not conduct electricity. Explain why.

_____ [1 mark]

d The formation of simple molecules containing single and double covalent bonds can be represented by dot and cross models.
  i Draw the dot and cross diagram of water.

[2 marks]

  ii Draw the dot and cross diagram of carbon dioxide.

[2 marks]

2 Carbon dioxide and water have very low melting points. Use the idea of **intermolecular forces** to explain why.

_____
_____ [2 marks]

3 a Sodium is in group 1. Explain why.

_____ [1 mark]

b Chlorine atoms have 7 electrons in the outer shell. In which group is chlorine?

_____ [1 mark]

c i To which period does fluorine belong?

_____ [1 mark]

  ii Explain why.

_____ [1 mark]

d The electronic structure of sulfur is 2, 8, 6. In which group is sulfur?

_____ [1 mark]

The group 1 elements

1 a Lithium, sodium and potassium react with water.

   i Which gas is given off?

   _____ [1 mark]

   ii They float on the surface. Explain why.

   _____ [1 mark]

b Sodium reacts very vigorously with water and forms sodium hydroxide. Write down the word equation for the reaction of sodium with water.

_____ [2 marks]

c Reactivity of the alkali metals with water increases down group 1.

reactivity increases down ↓

	melting point in °C	boiling point in °C
$_3$Li	179	1317
$_{11}$Na	98	
$_{19}$K		774

Estimate the melting point of potassium _____ and

the boiling point of sodium _____. [2 marks]

d Group 1 metals have similar properties. Explain why.

_____ [1 mark]

e Write a balanced symbol equation for the reaction of sodium metal with water.

_____ [2 marks]

2 a Marie and Mitch carried out some flame tests. How did they do this?

_____
_____
_____ [4 marks]

b Draw a **straight** line to match up their results.

red	potassium
yellow	lithium
lilac	sodium

[2 marks]

3 a Alkali metals have similar properties because when they react their atoms need to lose one electron to form full outer shells. Write down the equation to show the formation of a lithium ion from its atom.

_____ [1 mark]

b Sodium is more reactive than lithium. Explain why. Use ideas about numbers of shells of electrons.

_____
_____ [3 marks]

c What is the process of electron loss called?

_____ [1 mark]

The group 7 elements

1 a There is a **trend** in the **physical appearance** of the halogens at room temperature. Finish the table.

chlorine	
bromine	*orange liquid*
iodine	

[2 marks]

b Group 7 elements have similar properties. Explain why.

_____ [1 mark]

c Chlorine has an electronic structure of 2, 8, 7. It gains one electron to become 2, 8, 8.
  i Write an ionic equation to show this.

_____ [1 mark]

  ii What is the process of electron gain called?

_____ [1 mark]

  iii Fluorine is more reactive than chlorine. Explain why. Use ideas about the gain of electrons.

_____
_____ [2 marks]

2 a When a halogen reacts with an alkali metal a **metal halide** is made. Write down the word equation for the reaction between potassium and iodine.

_____ [2 marks]

b Potassium reacts with chlorine to produce potassium chloride. Write a balanced equation to show this reaction.

_____ [2 marks]

3 a If halogens are bubbled through **solutions of metal halides** there are two possibilities: **no reaction**, or a **displacement reaction**.
  i If chlorine is bubbled through potassium bromide solution a red-brown colour is seen. Explain why.

_____ [1 mark]

  ii If bromine is bubbled through potassium chloride solution there is no reaction. Explain why.

_____ [1 mark]

b Bromine displaces iodine from potassium iodide solution.
  i Write down a word equation for this reaction.

_____ [2 marks]

  ii Write down a balanced equation for this reaction.

_____ [2 marks]

C3 THE PERIODIC TABLE

Electrolysis

1 a Explain the key features of the electrolysis of dilute sulfuric acid.

_____

_____

_____

_____

_____

_____ [6 marks]

b Explain why the volume of hydrogen gas and the volume of oxygen gas given off in this process are different.

_____ [1 mark]

2 The electrolysis of sodium chloride takes place in solution. Describe the reactions at each of the electrodes. Explain the electron transfer processes in each case.

a At the cathode _____

_____

_____ [3 marks]

b At the anode _____

_____

_____ [3 marks]

3 a Write about the key features of the production of aluminium by electrolytic decomposition.

_____

_____

_____

_____ [4 marks]

b Write down the word equation for the decomposition of aluminium oxide.

_____ [1 mark]

c The electrode reactions in the electrolytic extraction of aluminium involve the transfer of electrons. Explain how reduction and oxidation occur at the electrodes.

_____

_____

_____

_____ [4 marks]

d Why is the chemical cryolite added to aluminium oxide?

_____

_____ [2 marks]

Transition elements

1 a A compound that contains a transition element is often coloured.
  i What is the colour of copper compounds?
  _____

  ii What is the colour of iron(II) compounds?
  _____

  iii What is the colour of iron(III) compounds?
  _____
  [3 marks]

b A transition metal and its compounds are often catalysts.
  i Which transition metal is used in the Haber process to produce ammonia?
  _____
  [1 mark]

  ii If the metal used to harden margarine is number 28, suggest whether this is a transition metal or not. Use the periodic table on page 4 to help you.
  _____
  [1 mark]

c If a transition metal carbonate is heated, it decomposes to form a metal oxide and carbon dioxide. Write down a word equation for the decomposition of copper carbonate.
  _____
  [1 mark]

2 Sodium hydroxide solution is used to identify the presence of transition metal ions in solution. Finish the table.

ion	colour
Cu^{2+}	
Fe^{2+}	
Fe^{3+}	

[3 marks]

3 a Copper carbonate decomposes into copper oxide and carbon dioxide. Write the balanced symbol equation for this thermal decomposition.
  _____
  [2 marks]

b Iron(III) ions react with hydroxide ions to from a precipitate. Write a balanced ionic symbol equation to show this precipitation reaction.
  _____
  [3 marks]

Metal structure and properties

1 a Silver is often chosen to make a piece of jewellery. Put a ring around the **two** properties that are important for this.

       ductile   good electrical conductor   high boiling point

       high melting point   lustrous   malleable

       good thermal conductivity   [2 marks]

b Copper is often used for the base or the whole of a saucepan. Use your knowledge about chemical properties to explain why.

_____

_____

_____ [3 marks]

c Aluminium is used in the aircraft industry and in modern cars. Explain why.

_____ [1 mark]

d Label the diagram to explain a **metallic bond**.

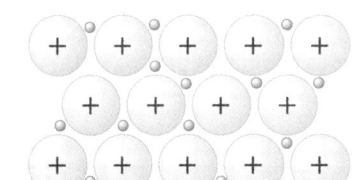

[1 mark]

e Metals have high melting points and boiling points. Use ideas about delocalised electrons to explain why.

_____

_____ [3 marks]

2 a What are superconductors?

_____ [1 mark]

b What are **two** potential benefits of superconductors?

_____

_____ [2 marks]

c Why do scientists need to develop superconductors that will work at 20 °C?

_____ [1 mark]

d Explain why metals conduct electricity.

_____ [1 mark]

e

metal	electrical conductivity 10^6/cm Ω	thermal conductivity W m^{-1} K^{-1}	tensile strength MPa
A	0.099	80	350
B	0.452	320	170
C	0.596	400	210

  i Which metal would be used for electrical cables? _____ [1 mark]

  ii Which metal would be used for pulling up elevators (lifts)? _____ [1 mark]

C3 Revision checklist

- I know that the nucleus is made up of protons and neutrons, with each having a relative mass of 1. ☐

- I know that electrons surround the nucleus and occupy shells in order. They have almost no mass, 0. ☐

- I know that positive ions are formed by the loss of electrons from the outer shell. ☐

- I know that negative ions are formed by the gain of electrons into the outer shell. ☐

- I know that non-metals combine by sharing electrons, which is called covalent bonding. ☐

- I can draw the 'dot and cross' diagrams of simple molecules such as H_2, Cl_2, CO_2 and H_2O. ☐

- I know that lithium, sodium and potassium react vigorously with water and give off hydrogen. ☐

- I know that group 1 metals have one electron in their outer shell, which is why they are similar. ☐

- I know that chlorine is a green gas, bromine is an orange liquid and iodine is a grey solid. ☐

- I know that halogens gain one electron to form a stable outer shell. This is called reduction. ☐

- I know that in the electrolysis of dilute sulfuric acid, H_2 is made at the cathode and O_2 at the anode. ☐

- I know that when aluminium oxide is electrolysed, Al is formed at the cathode and O_2 at the anode. ☐

- I know that compounds of copper are blue, iron(II) are light green and iron(III) are orange/brown. ☐

- I know that metals conduct electricity as the 'sea' of electrons move through the positive metal ions. ☐

C3 THE PERIODIC TABLE

Acids and bases

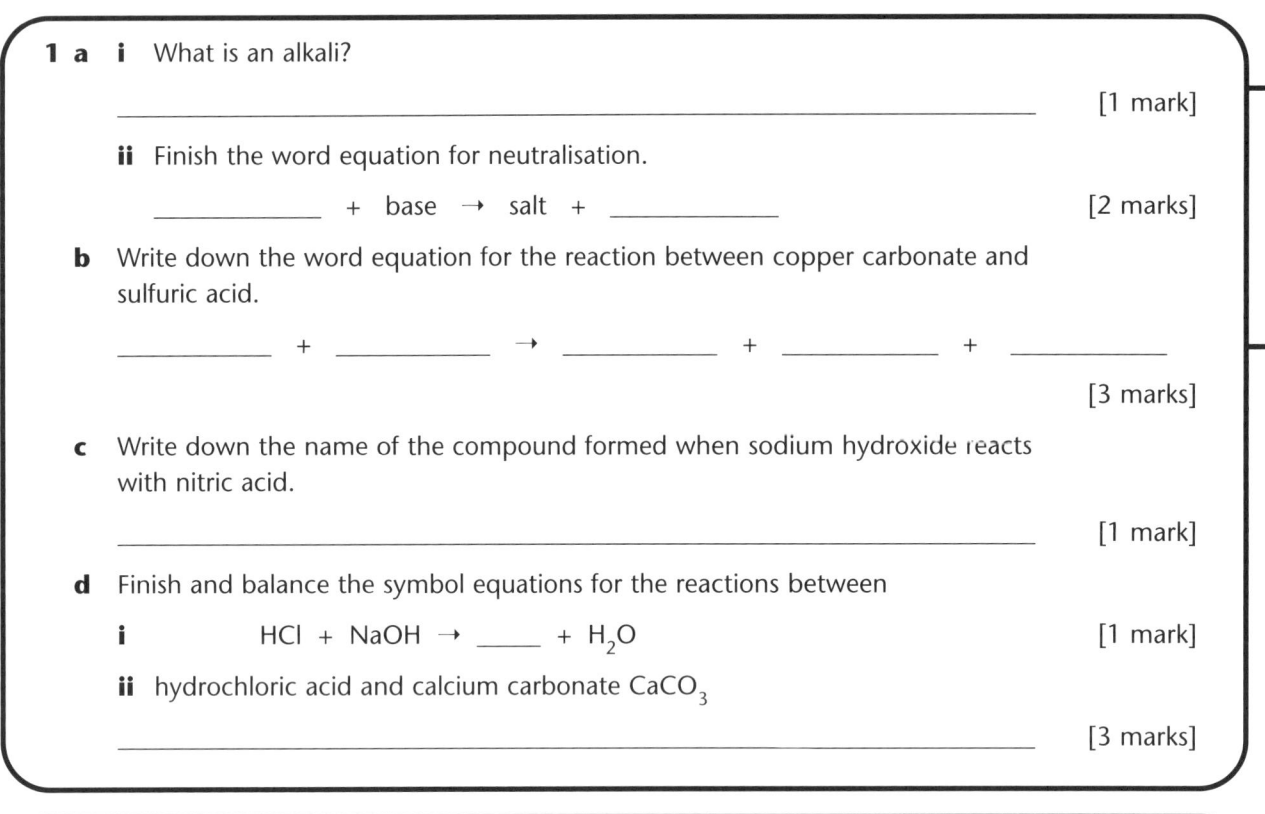

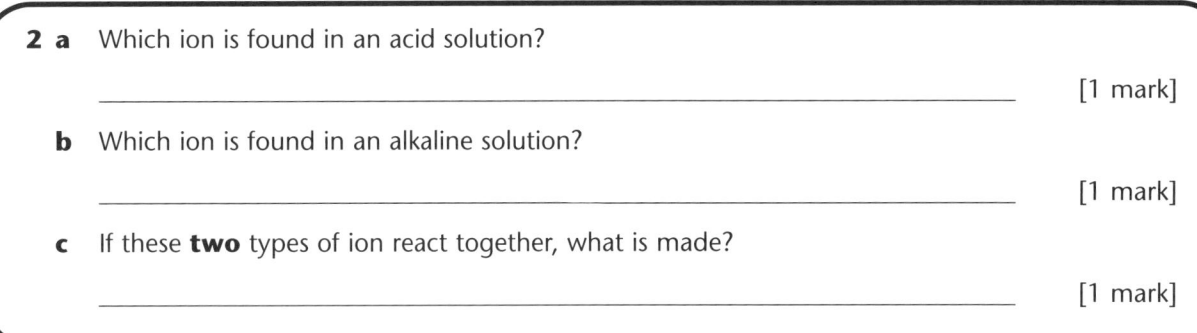

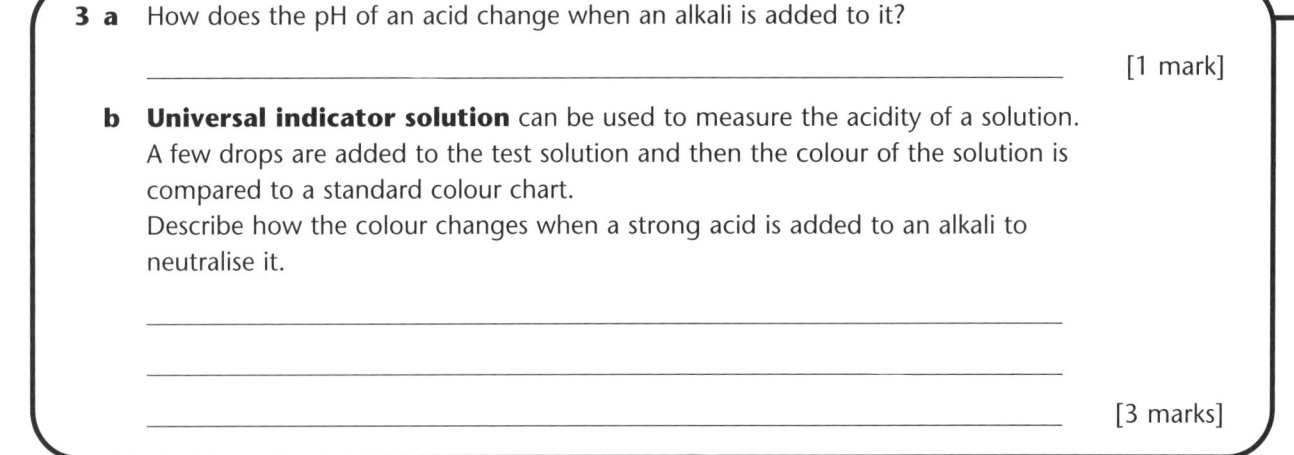

1 a i What is an alkali?

_____ [1 mark]

   ii Finish the word equation for neutralisation.

      _____ + base → salt + _____ [2 marks]

b Write down the word equation for the reaction between copper carbonate and sulfuric acid.

_____ + _____ → _____ + _____ + _____

[3 marks]

c Write down the name of the compound formed when sodium hydroxide reacts with nitric acid.

_____ [1 mark]

d Finish and balance the symbol equations for the reactions between

  i   HCl + NaOH → _____ + H_2O [1 mark]

  ii hydrochloric acid and calcium carbonate $CaCO_3$

_____ [3 marks]

2 a Which ion is found in an acid solution?

_____ [1 mark]

b Which ion is found in an alkaline solution?

_____ [1 mark]

c If these **two** types of ion react together, what is made?

_____ [1 mark]

3 a How does the pH of an acid change when an alkali is added to it?

_____ [1 mark]

b **Universal indicator solution** can be used to measure the acidity of a solution. A few drops are added to the test solution and then the colour of the solution is compared to a standard colour chart.
Describe how the colour changes when a strong acid is added to an alkali to neutralise it.

_____

_____

_____ [3 marks]

C4 CHEMICAL ECONOMICS

Grades: D–C, B–A*, B–A*, D–C

Reacting masses

1 a Work out the relative formula mass of NaOH.

_____ [1 mark]

b Work out the relative formula mass of CaCO$_3$.

_____ [1 mark]

c Work out the relative formula mass of Ca(OH)$_2$.
Use the relative atomic masses.

H 1   C 12   O 16   Na 23   Ca 40

_____ [1 mark]

d Explain why mass is conserved in a chemical reaction.

_____
_____ [2 marks]

2 a Leo and Lesley made some crystals of magnesium sulphate. They did not make as much as they hoped. They wanted to make 42 g. They only made 28 g.

i What was their 'actual yield'?

_____ [1 mark]

ii What was their 'predicted yield'?

_____ [1 mark]

iii How will they calculate their percentage yield?

_____
_____ [1 mark]

iv What was their percentage yield?

_____
_____ [2 marks]

b Zinc carbonate decomposes on heating to give zinc oxide and carbon dioxide.

i Write a balanced symbol equation for this reaction.

_____ [2 marks]

ii Calculate how much CO$_2$ is made when 12.50 g of ZnCO$_3$ decomposes.

_____
_____
_____
_____ [4 marks]

C4 CHEMICAL ECONOMICS

Fertilisers and crop yield

C4 CHEMICAL ECONOMICS

Grades

1 a Why do farmers add fertilisers to their crops?
_____ [1 mark] — D–C

b How do fertilisers get into the plants?
_____ [1 mark]

c Why is nitrogen needed for increased plant growth?
_____ [1 mark] — B–A*

2 a To calculate the yield when making a fertiliser you need to calculate its **relative formula mass**. What is the relative formula mass of ammonium sulphate $(NH_4)_2SO_4$? Use the periodic table on page 4.
_____ [2 marks] — D–C

b Farmers can use relative formula masses to find the percentage of each element in a fertiliser – it is printed on the bag for them. They use the formula:

percentage of element = $\dfrac{\text{mass of the element in the formula}}{\text{relative formula mass}} \times 100$

Calculate the percentage of nitrogen in potassium nitrate, KNO_3.
_____
_____
_____ [3 marks] — B–A*

3 a Many fertilisers are **salts**, so they can be made by reacting acids with bases. What else is made?

acid + base → salt + _____ [1 mark] — D–C

b Don and Demi want to make some ammonium phosphate.

  i Which acid will they need to use? _____ [1 mark]

  ii Which alkali will they need to use? _____ [1 mark]

  iii Write down a word equation to show this reaction.
_____ [2 marks]

c Describe in steps how Don and Demi make ammonium phosphate from their acid and alkali.
_____
_____
_____
_____ [6 marks] — B–A*

4 Too much fertilisers applied may increase the level of nitrates or phosphates in a water course. This may cause eutrophication to occur. Explain how this happens. — B–A*
_____
_____
_____ [3 marks]

The Haber process

1 a Write about how ammonia is made. Include the conditions needed in your answer.

_____
_____
_____ [3 marks]

b As the reaction for producing ammonia is reversible, the percentage yield for the reaction cannot be 100%.

i How does the pressure affect the reaction?

_____
_____ [2 marks]

ii How does the temperature affect the reaction?

_____
_____ [2 marks]

iii Give the optimum temperature for this reaction and explain why it is used.

_____
_____
_____ [2 marks]

2 a Look at the graph. At which pressure is most ammonia made at 400 °C?

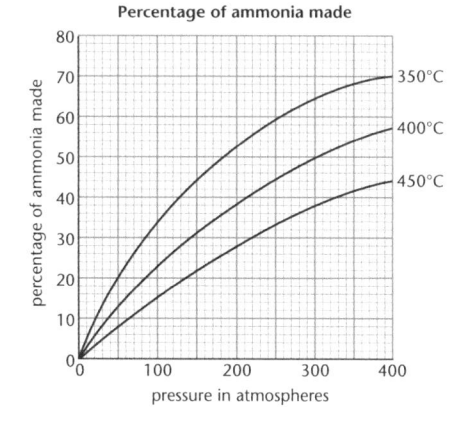

Percentage of ammonia made

_____ [1 mark]

b As the pressure increases what happens to the yield of ammonia?

_____ [1 mark]

c As the temperature increases what happens to the yield of ammonia?

_____ [1 mark]

d Chemical plants need to work at the conditions that produce the highest percentage yield for a reaction, most cheaply. Use ideas about optimum conditions to explain how this is achieved in the manufacture of ammonia. Explain why these conditions cut costs.

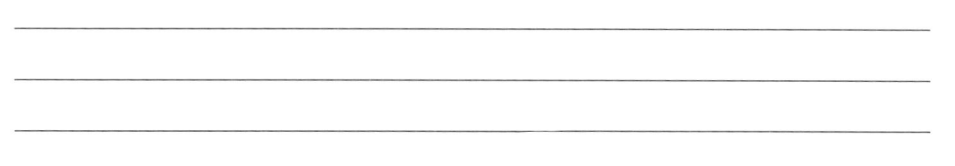

_____
_____
_____
_____
_____ [5 marks]

Detergents

1 a A detergent can be made by **neutralising** an organic acid using an alkali. Write a word equation for this reaction.

_____ [2 marks]

b Why are detergents used to clean greasy plates?

_____
_____ [2 marks]

c New washing powders allow clothes to be washed at low temperatures.
   i This is good for the environment. Explain why.

_____
_____ [3 marks]

   ii It is also good for coloured clothes to be washed at low temperatures. Explain why.

_____
_____ [2 marks]

d When clothes are washed, grease is lifted off the clothes and put into water by detergent.
Use the words **hydrophilic** and **hydrophobic** to explain the stages in the diagrams.

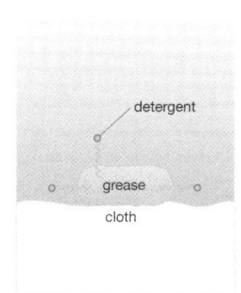

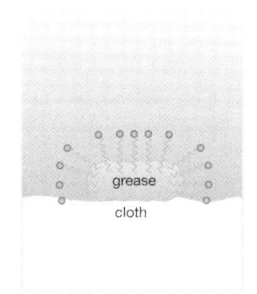

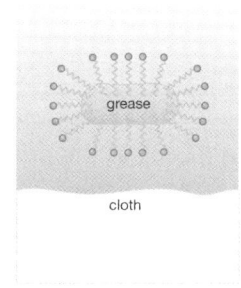

[3 marks]

2 a The forces that hold molecules of grease together and molecules of dry-cleaning solvent together are forces between molecules. What are these are called?

_____ [1 mark]

b Molecules of water are held together by stronger forces called **hydrogen bonds**. The water molecules cannot stick to the grease because they are sticking to each other much too strongly. On the diagram mark the **covalent** bonds between atoms and the **hydrogen** bonds between water molecules.

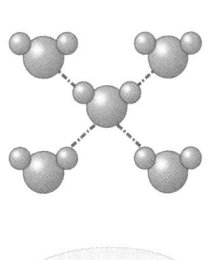

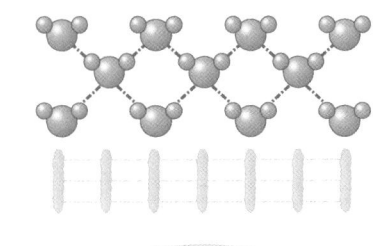

[5 marks]

Batch or continuous?

1 a Why are pharmaceuticals made in small batches?

_____ [1 mark]

b How is the large scale production of ammonia different from the small scale production of pharmaceuticals?

_____ [1 mark]

c A **continuous** process plant is effective. Explain why.

_____
_____
_____
_____ [4 marks]

d **Batch** processes are not as efficient but have one major advantage. Explain what this is.

_____ [1 mark]

2 Draw **straight** lines to match the reasons for the high costs of making and developing medicine and pharmaceutical drugs to the **best** explanation.

strict safety laws	The medicines are made by a batch process so less automation can be used.
research and development	They may be rare and costly.
raw materials	They take years to develop.
labour intensive	People need to feel a benefit without too many side effects.

[3 marks]

3 Whether or not a drug is developed depends on a number of economic considerations. Explain what these are.

_____
_____
_____
_____
_____ [6 marks]

Nanochemistry

1 a Draw a **straight** line to match the carbon to its correct structure.

diamond

graphite

buckminster fullerene

[3 marks]

b Finish the table to show **one** use of the carbon type and a reason for that use.

	diamond	graphite	buckminster fullerene
use			semiconductors in electrical circuits
reason			good electrical conductor on small scale

[4 marks]

c Explain why these structures have these properties.

diamond	graphite
Does not conduct electricity because _____	Slippery because _____

[2 marks]

2 a What are different forms of the same element called?

_____ [1 mark]

b Explain why the giant structure of diamond is so hard and does not conduct electricity.

_____

[5 marks]

3 a Write down **two** uses of nanotubes.

_____

[2 marks]

b Why are nanotubes so useful as catalysts?

_____ [1 mark]

How pure is our water?

1 a The water in a river is cloudy and often not fit to drink. To make clean drinking water it is passed through a **water purification** works.
Label the **three** main parts of this process.

☐ → ☐ → ☐

[3 marks]

b Explain the stages of water purification in detail.

_____
_____
_____
_____

[3 marks]

c i Seawater has many substances dissolved in it so it is undrinkable. Which special technique has to be used to remove these unwanted substances?

_____

[1 mark]

ii Why is this technique not used for all purification of water?

_____
_____

[1 mark]

2 a Explain why relief organisations concentrate on providing clean water supplies.

_____
_____

[2 marks]

b Water is a **renewable resource** but not an endless resource in any one country. Explain why.

_____
_____

[2 marks]

3 a Write a word equation for the precipitation reaction between lead nitrate and potassium chloride.

_____

[2 marks]

b Write a word equation for the precipitation reaction between silver nitrate $AgNO_{3(aq)}$ and potassium bromide $KBr_{(aq)}$.

_____

[1 mark]

c Write a full balanced symbol equation for this precipitation reaction.

_____

[3 marks]

C4 CHEMICAL ECONOMICS

C4 Revision checklist

C4 CHEMICAL ECONOMICS

- I know that neutralisation is a reaction where: acid + base → salt + water. ☐

- I can construct balanced symbol equations, such as:
  $2KOH + H_2SO_4 \rightarrow K_2SO_4 + H_2O$ ☐

- I can work out the relative formula mass of a substance from its formula, such as $Ca(OH)_2$. ☐

- I can work out the percentage yield using the formula:
  % yield = actual yield × 100 ÷ predicted yield. ☐

- I know that fertilisers provide extra essential elements but excess can cause eutrophication. ☐

- I know that ammonia is made by the Haber process where N_2 and H_2 are put over an iron catalyst. ☐

- I know that the higher the pressure, the higher the percentage yield of ammonia. ☐

- I know that a catalyst will increase the rate of reaction but will not change the percentage yield. ☐

- I know that a detergent has a hydrophilic head and a hydrophobic tail. ☐

- I know that dry cleaning is a process used to clean clothes using a solvent that is not water. ☐

- I know that a continuous process makes chemicals all the time but a batch process does not. ☐

- I can recognise the three allotropes of carbon: diamond, graphite and buckminster fullerene. ☐

- I can explain that graphite is slippery and is used as electrodes as it conducts electricity. ☐

- I know that water purification includes filtration, sedimentation and chlorination. ☐

Moles and empirical formulae

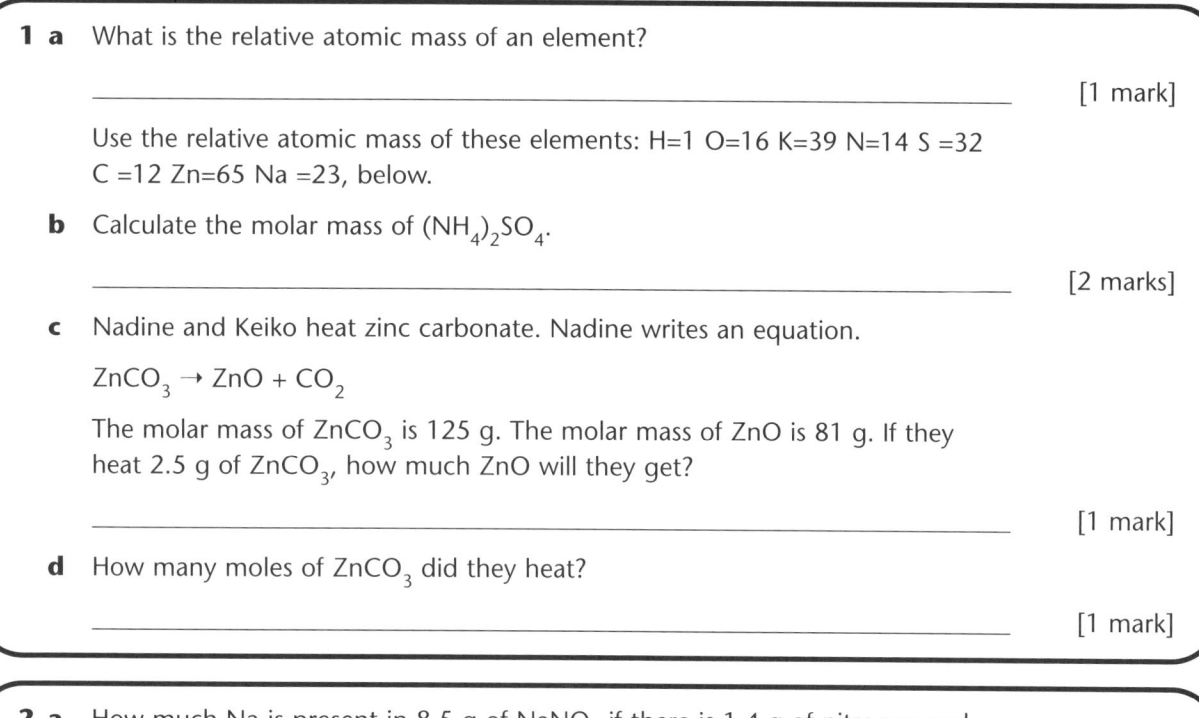

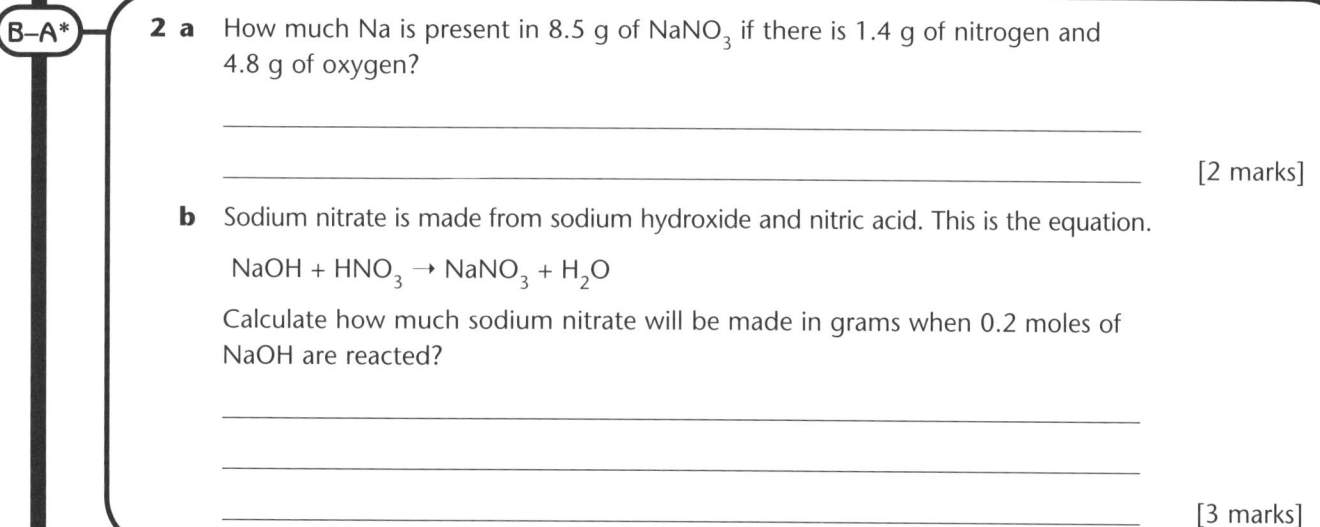

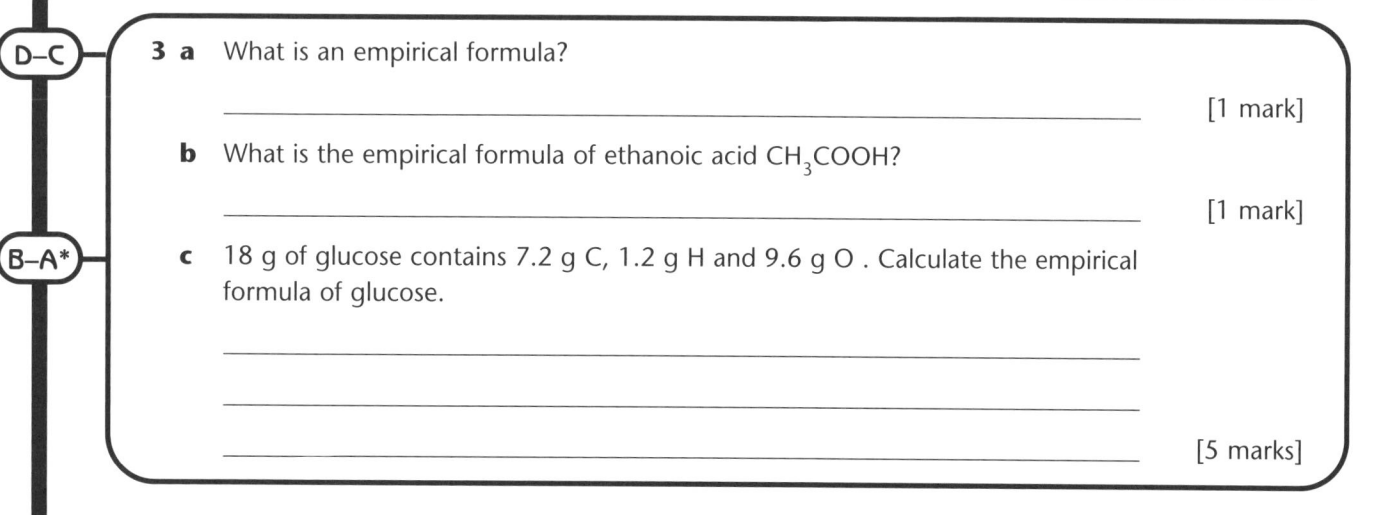

1 a What is the relative atomic mass of an element?

_____ [1 mark]

Use the relative atomic mass of these elements: H=1 O=16 K=39 N=14 S=32 C=12 Zn=65 Na=23, below.

b Calculate the molar mass of $(NH_4)_2SO_4$.

_____ [2 marks]

c Nadine and Keiko heat zinc carbonate. Nadine writes an equation.

$ZnCO_3 \rightarrow ZnO + CO_2$

The molar mass of $ZnCO_3$ is 125 g. The molar mass of ZnO is 81 g. If they heat 2.5 g of $ZnCO_3$, how much ZnO will they get?

_____ [1 mark]

d How many moles of $ZnCO_3$ did they heat?

_____ [1 mark]

2 a How much Na is present in 8.5 g of $NaNO_3$ if there is 1.4 g of nitrogen and 4.8 g of oxygen?

_____
_____ [2 marks]

b Sodium nitrate is made from sodium hydroxide and nitric acid. This is the equation.

$NaOH + HNO_3 \rightarrow NaNO_3 + H_2O$

Calculate how much sodium nitrate will be made in grams when 0.2 moles of NaOH are reacted?

_____
_____
_____ [3 marks]

3 a What is an empirical formula?

_____ [1 mark]

b What is the empirical formula of ethanoic acid CH_3COOH?

_____ [1 mark]

c 18 g of glucose contains 7.2 g C, 1.2 g H and 9.6 g O. Calculate the empirical formula of glucose.

_____
_____
_____ [5 marks]

Electrolysis

1 a Lukas and Anya electrolyse a solution of potassium nitrate in water. They collect two gases, one at each electrode. Which two gases?

_____ and _____ [2 marks]

b Lukas says that electrolysis is the decomposition of a liquid by passing an electric current through it. Anya says there is a more detailed explanation. What does she say?

_____ [2 marks]

c i Anya wants to electrolyse potassium nitrate solution to get **potassium**. Lukas says she will **only** get the same gases at each electrode as in their previous experiment. Explain why.

_____ [1 mark]

ii Lukas explains further by writing the half equation for the processes that happen at each electrode in the electrolysis of KNO_3 in water in a table. Ions present: H^+, K^+, NO_3^-, OH^-.

He uses e^- for an electron. Complete his table.

at the cathode	at the anode

[4 marks]

2 Lukas and Ben electrolyse copper sulfate. They use copper electrodes. They find the mass of the electrodes before and after the electrolysis.

a i What do they notice about the change in mass of the negative electrode?

_____ [1 mark]

ii What do they notice about the change in mass of the positive electrode?

_____ [1 mark]

b Lukas says that if they leave the electrolysis for longer they will get more copper. Ben says if they increase the current they will get more copper. Who is correct? Put a (ring) round the correct answer.

Ben   **Both**   **Lukas**   **Neither** [1 mark]

c They noted a mass increase on one electrode of 0.36 g. They were using 0.1 amps for 1.5 hours. Work out what mass would they get if they transferred 270 coulombs instead?

_____
_____
_____
_____ [4 marks]

3 a Anya and Jenny write half equations of what they would get at each electrode with these molten electrolytes. Fill in their table.

molten electrolyte	at the cathode	at the anode
Al_2O_3		
$PbBr_2$		

[4 marks]

b Why do these electrolytes need to be molten?

_____ [1 mark]

Quantitative analysis

1 Sean and Mary wanted to eat some pie. They look on the label.

	Pie per 100 g	% RDA	Oatcake per 100 g
Energy	860 kJ	10%	1837 kJ
Protein	5.0		10.4 g
Carbohydrate:	19.0 g		57.4 g
of which sugar	2.7 g	3%	1.8 g
Fat	12.2 g	17%	18.5 g
of which saturates	5.0 g	25%	2.7 g
of which unsaturates	7.2 g		15.8 g
Fibre	1.0 g		8.0 g
Sodium	0.1 g		0.8 g
Sodium equivalent	0.3 g	5%	

a If they eat 200 g of pie, how much saturated fat would they eat?

_____ [1 mark]

b If they eat 200 g of pie, what percentage of their RDA (recommended daily allowance) of saturated fat will they eat?

_____ [1 mark]

c Mary says they need to eat fibre. There are 8 oatcakes in 100 g. She compares the fibre content of 100 g of pie and 1 oatcake. What does she notice?

_____ [2 marks]

d Work out the salt equivalent of 100 g of oatcakes.

_____ [1 mark]

2 a Mary finds that a squash bottle holds 750 cm³. How many dm³ is this?

_____ [1 mark]

b The squash is too concentrated to drink. It has 15 g/dm³. She pours 10 cm³ of it into a glass. How much water must Mary add to make the squash into a 5 g/dm³ solution?

_____ [2 marks]

c i A solution of potassium hydroxide has 56 g/dm³. How many moles/dm³ is this? (Relative atomic mass: K =39, O =16, H=1, Na =23)

_____ [1 mark]

ii How many g/dm³ are needed to make a 0.1 mol/dm³ solution of NaOH?

_____ [1 mark]

iii Calculate the concentration of a solution of NaOH in mol/dm³ that has 0.8 g NaOH in 1000 cm³.

_____ [2 marks]

3 a Sean says that solutions are made up of particles. How would he explain a concentrated solution?

_____ [1 mark]

b Mary knows that to dilute an acid she must add the acid to the water and not the other way round. How would she dilute a solution of hydrochloric acid of 1 mol/dm³ to make 100 cm³ of a solution of 0.1 mol/dm³?

_____ [2 marks]

Titrations

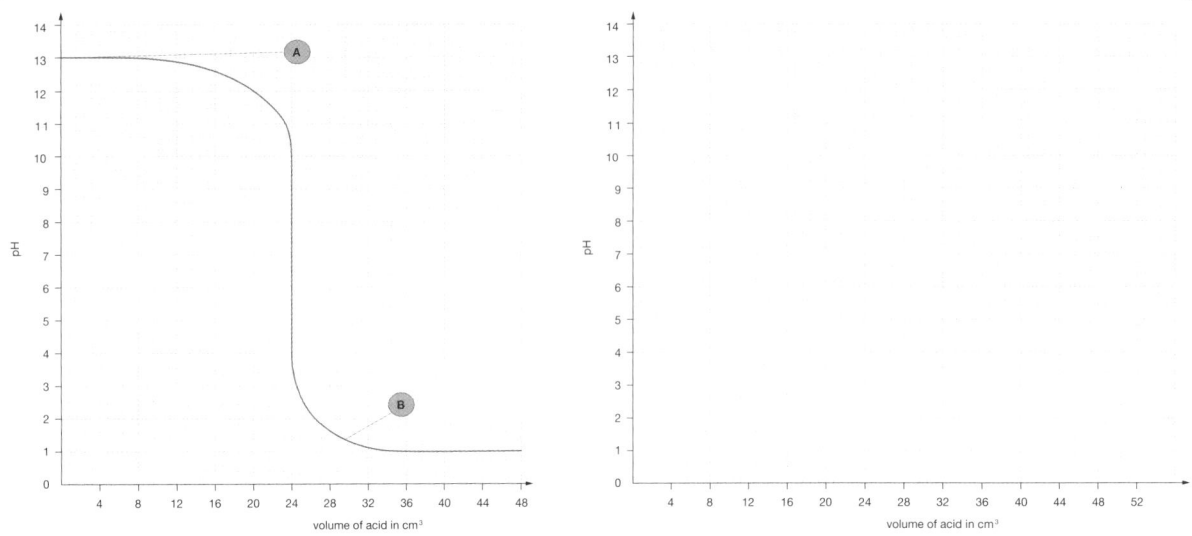

1 a Look at the pH curve. It follows the reaction between an acid and an alkali. Write down what is happening at A and B.

   A _____

   B _____ [2 marks]

b When does the end point of this reaction happen? _____ [1 mark]

c What is the pH when 28 cm³ acid have been added? _____ [1 mark]

d The pH changes during the reaction of an acid and an alkali. Explain why.

   _____
   _____ [2 marks]

e On the second grid sketch another pH curve. The reaction is between an alkali pH 13.5 and an acid pH 1.4. The alkali neutralises exactly 23.5 cm³ acid. [2 marks]

2 a Chen-chi is measuring the reaction between an acid and an alkali. She carries out several titrations. Explain why.

   _____ [1 mark]

b She finds that 23.1 cm³ of hydrochloric acid exactly neutralises 25 cm³ of sodium hydroxide solution. The concentration of the hydrochloric acid solution is 0.12 mol/dm³. She knows that 1 mole of HCl neutralises 1 mole of NaOH. She calculates the concentration of NaOH solution in mol/dm³. Show how.

   _____
   _____
   _____ [3 marks]

3 a Indicators are used in titrations. What does a single indicator, like litmus, do that is different to Universal indicator when used in a titration.

   _____
   _____ [2 marks]

b Litmus is used in preference to Universal indicator in a titration. Explain why.

   _____
   _____ [2 marks]

Gas volumes

1 a Asif measures the amount of gas given off when marble chips react with acid.

Describe how he does this, using a mass balance.

_____

_____

_____ [3 marks]

b Haleema measures the amount of gas given off when magnesium reacts with acid.

She does this using a gas syringe.

Describe how.

_____

_____

_____ [3 marks]

2 a Haleema drew a graph of her results, labelled reaction 1. She repeated the reaction but this time with only half the amount of magnesium. These results are reaction 2.

   i What is the total volume of gas produced in reaction 1?

_____[1 mark]

   ii At what time did reaction 1 stop? Why did it stop?

_____

_____ [2 marks]

   iii In reaction 1 how much gas had she collected at 12 seconds?

_____ [1 mark]

   iv Draw on the graph to find the rate of reaction in the first 5 seconds. [3 marks]

b i Sketch the line that Haleema would get for the changing volume of gas in reaction **2**. [1 mark]

   ii Explain why Haleema would predict that final volume of gas sketched for reaction 2.

_____ [1 mark]

c Next Haleema used 0.12 g of magnesium with excess acid. What volume of gas should she get from the reaction: Mg + 2HCl → $MgCl_2$ + H_2 (Relative atomic mass: Mg = 24. One mole of gas occupies 24 dm^3 at room temperature and pressure.)

_____

_____

_____ [3 marks]

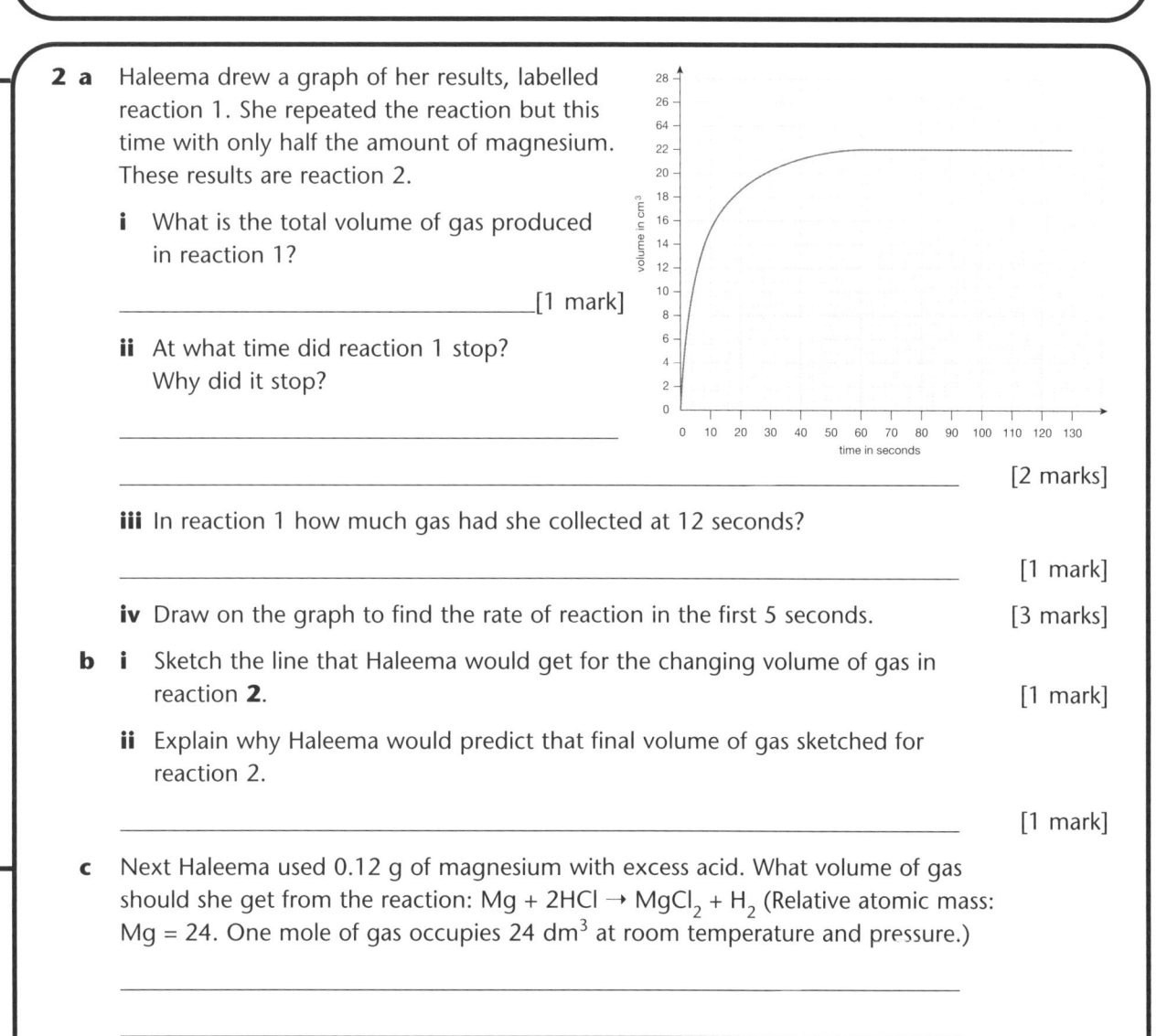

Equilibria

1 a Some reactions reach equilibrium. If A + B ⇌ C reaches equilibrium, what does this tell us about the rate of the reactions?

_____ [1 mark]

b If the concentration of C is greater than the concentration of A + B in the reaction A + B ⇌ C, what does this tell us about where the equilibrium lies?

_____ [1 mark]

c Explain why reversible reactions reach equilibrium. Give at least two steps.

_____
_____
_____ [3 marks]

2 a The percentage of ammonia produced changes as the pressure and temperature of the reaction changes. Look at the graph.

i What is the percentage of ammonia made at 270 atmospheres and 350 °C?

_____ [1 mark]

ii How does the percentage of ammonia change as the **pressure** increases?

_____ [1 mark]

iii How does the percentage of ammonia change as the **temperature** increases?

_____ [1 mark]

b What happens to the position of the equilibrium if the product is being removed?

_____ [1 mark]

c In the reaction $N_2 + 3H_2 \rightleftharpoons 2NH_3$, increasing the pressure moves the equilibrium to the right. Explain why. Use ideas about numbers of moles.

_____ [1 mark]

3 a The Contact Process is sulfur dioxide + oxygen ⇌ sulfur trioxide.

i Where does the sulfur dioxide come from?

_____ [1 mark]

ii Write down **three** symbol equations for the stages in the manufacture of sulfuric acid.

_____
_____
_____ [4 marks]

b What are the **three** conditions used in the Contact Process? Explain why they are used.

_____
_____
_____ [3 marks]

Strong and weak acids

1 Julie compares two acids A and B. They both have the same concentration. B is a strong acid, A is a weak acid.

  a Both acids ionise in water. Which ion is produced that makes the solution an acid?

    _____ [1 mark]

  b Both acids ionise to produce the same ion. Write down **how** B is a strong acid.

    _____ [2 marks]

  c When A ionises, an equilibrium mixture is made. What **type** of reaction is this ionisation?

    _____ [1 mark]

  d If B is hydrochloric acid, HCl, write an equation to show how it ionises.

    _____

    _____ [2 marks]

  e If A is ethanoic acid, CH_3COOH, write an equation to show how it ionises.

    _____ [3 marks]

  f The strength of hydrochloric acid is higher than the strength of ethanoic acid of the same concentration. Explain what is meant by:

    strength _____

    concentration _____ [2 marks]

  g Explain how the pH of a weak acid differs from the pH of a strong acid of the same concentration. _____ [1 mark]

2 Ben reacts hydrochloric acid and Gita reacts ethanoic acid with magnesium. They each used the same amount of acid and the same amount of magnesium. They both measured the volume of gas they got every 10 seconds.

Ben plotted his results as line C. Gita plotted her results as line D.

They use the same **concentration** of acid. Ben notices his acid reacts **faster** with magnesium and Gita notices that her acid reacts more **slowly**.

Explain why ethanoic acid reacts slower than hydrochloric acid with magnesium. Use ideas about ions, collisions and acid strength.

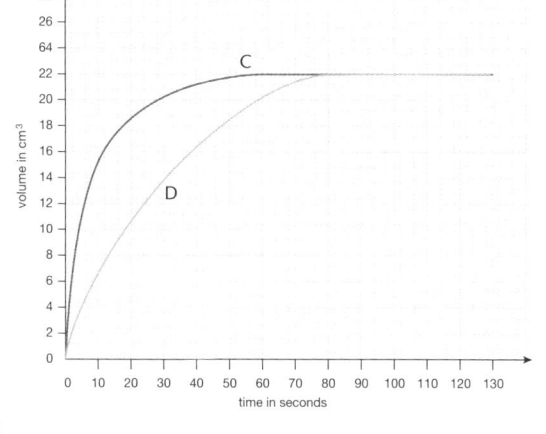

_____

_____

_____ [3 marks]

3 a Ethanoic acid has a lower electrical conductivity than hydrochloric acid. Explain why.

    _____ [2 marks]

  b If either acid is electrolysed, hydrogen is given off at the negative electrode. Explain why.

    _____ [2 marks]

Ionic equations

1 a Look at the diagram of a solid substance.

What happens to the particles when the substance is heated or put into solution?

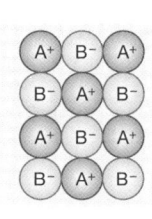

_____ [1 mark]

b Two solutions, silver nitrate and sodium bromide, react in a precipitation reaction. Describe how this happens and describe the speed of the reaction. Use ideas about ions.

_____
_____ [2 marks]

2 a Write a word equation for the reaction between silver nitrate and sodium bromide.

_____ [2 marks]

b Write a word equation for the test for chloride ions with silver nitrate. Use a sodium salt.

[ ] [ ] → [ ] [ ] [2 marks]

c Write a word equation for testing a sulfate using barium chloride. Use a potassium salt.

_____ [2 marks]

d Write an ionic equation for the reaction between silver nitrate (Ag^+ ions and NO^{3-} ions) and potassium iodide (K^+ and I^- ions) to make a precipitate of silver iodide.

_____ [3 marks]

e Jo reacts lead nitrate and sodium iodide solutions. Construct a balanced ionic equation for the reaction which makes a precipitate of lead iodide. Write in state symbols also. Formula of ions Na^+, Pb^{2+}, I^-, NO_3^-.

_____ [3 marks]

3 a Chris and Sam make lead iodide. They use these pieces of apparatus.

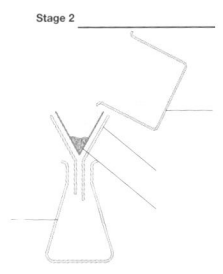

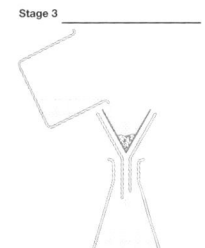

Stage 1 _____ Stage 2 _____ Stage 3 _____ Stage 4 _____

Complete the diagram by writing what Chris and Sam do at each stage. [4 marks]

b i When Chris and Sam make lead iodide. They write the balanced equation.

$Pb(NO_3)_2 + 2KI \rightarrow PbI + 2KNO_3$

Write out the ionic equation for this reaction. Give all the ions.

_____ [3 marks]

ii In this reaction there is a **spectator ion**. Explain what is meant by a spectator ion.

_____ [1 mark]

C5 Revision checklist

- I know that the empirical formula of glucose, $C_6H_{12}O_6$, is CH_2O ☐

- I know that the number of moles of a substance = mass ÷ molar mass ☐

- I know that the amount made during electrolysis increases if time or current increases ☐

- I know that the reaction at the cathode in the electrolysis of $CuSO_4$ with copper electrodes is $Cu^{2+} + 2e^- \rightarrow Cu$ ☐

- I know that the more concentrated a solution, the more crowded the particles. ☐

- I know that a 0.2 mol/dm³ solution of NaOH is 0.2 × 40 = 8 g/dm³ concentrated. ☐

- I know that an acid added to an alkali produces a salt and water. ☐

- I know how to sketch a pH titration curve for the titration of an acid with an alkali. ☐

- I know how to do an experiment to measure the volume of gas given off. ☐

- I know that 1 mole of gas occupies a volume of 24 dm³ at room temperature and pressure. ☐

- I know that the conditions needed for the Contact Process are V_2O_5 catalyst, temperature of 450 °C and atmospheric pressure. ☐

- I know that a catalyst increases the rate of a reaction but does not change the position of equilibrium. ☐

- I know that an acid ionises in water to produce H^+ ions. ☐

- I know that the strength of an acid depends on the degree of ionisation of the acid. ☐

- I know that the word equation for the test for sulfate is
barium chloride + sodium sulfate → barium sulfate + sodium chloride. ☐

- I know that a 'spectator ion' does not take part in the formation of a precipitate. ☐

Energy transfers – fuel cells

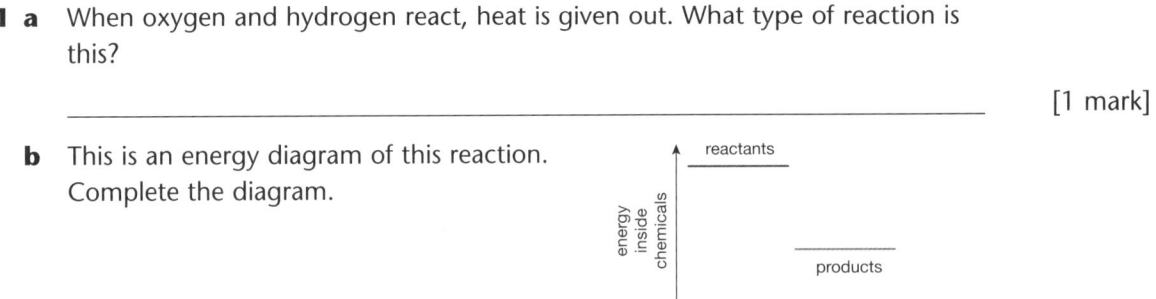

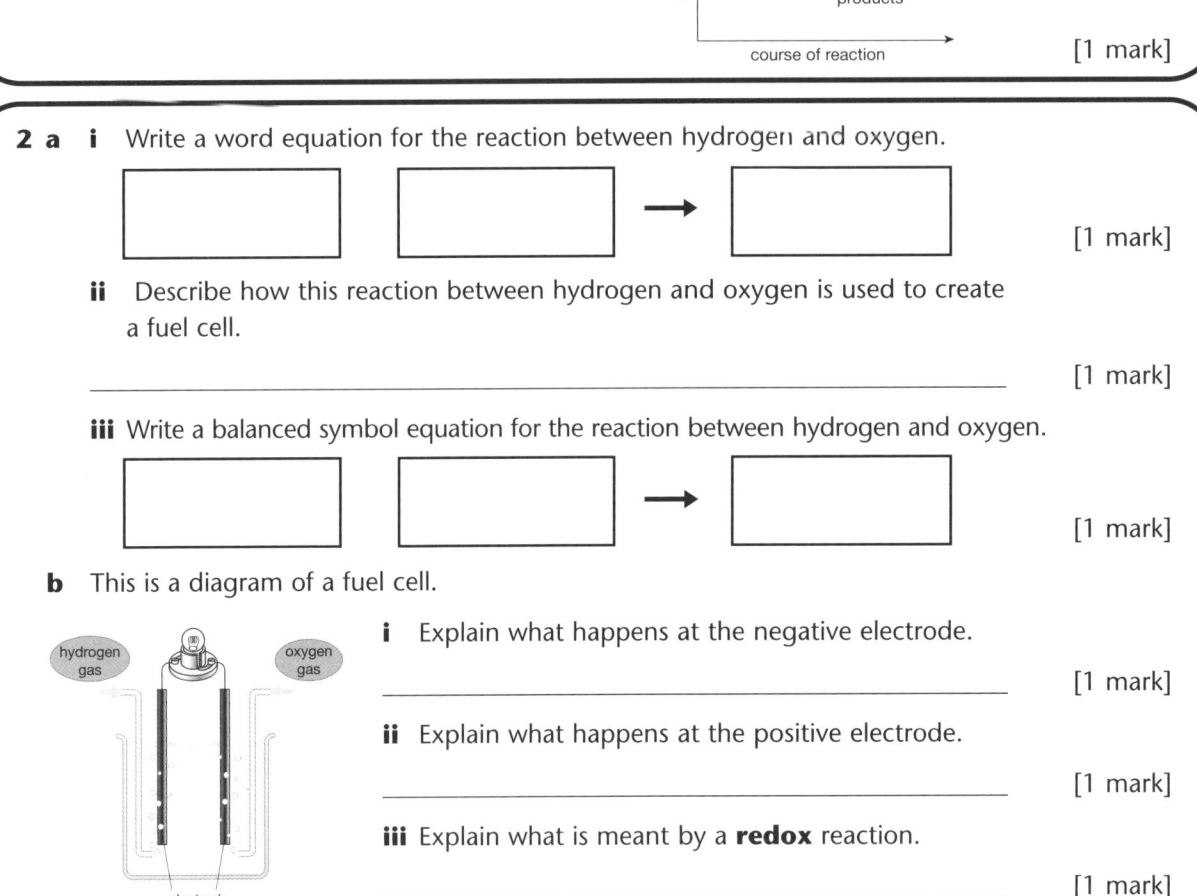

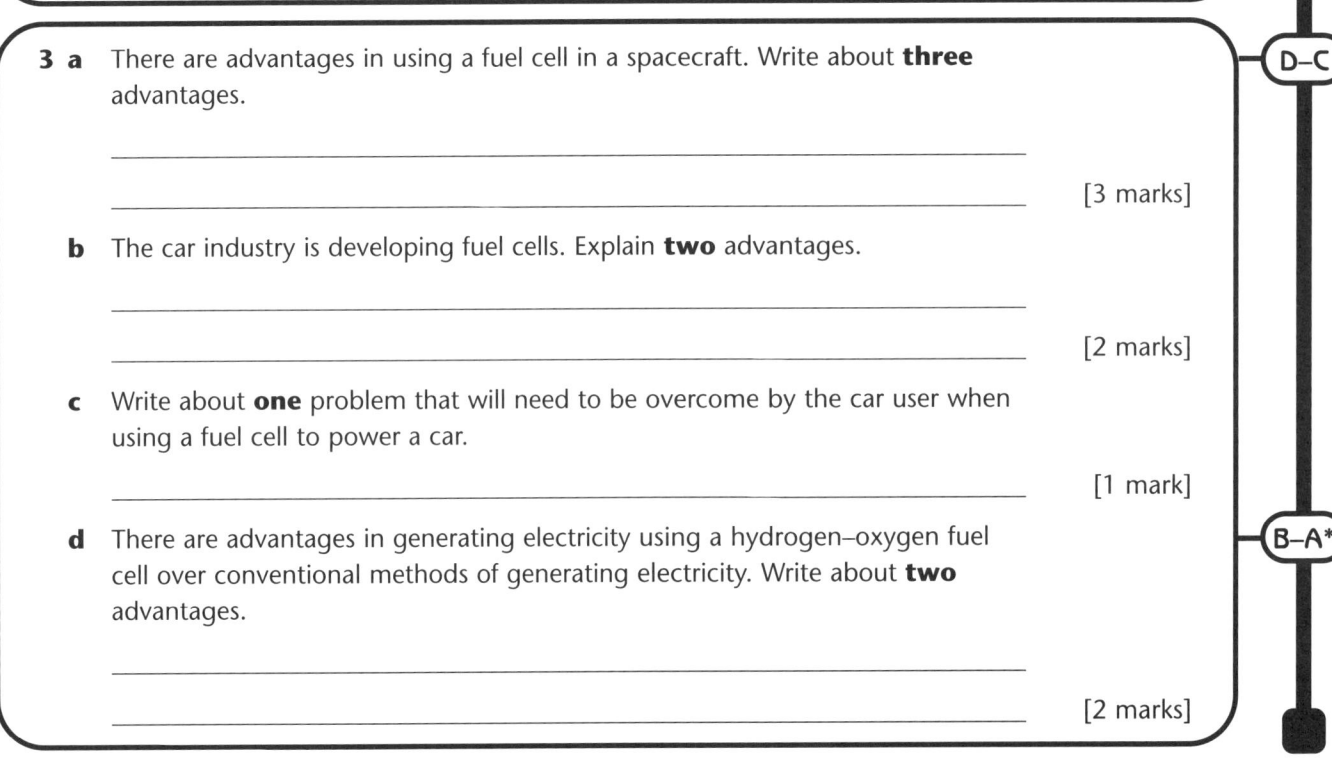

1 a When oxygen and hydrogen react, heat is given out. What type of reaction is this?

_____ [1 mark]

b This is an energy diagram of this reaction. Complete the diagram.

[1 mark]

2 a i Write a word equation for the reaction between hydrogen and oxygen.

☐ ☐ → ☐ [1 mark]

ii Describe how this reaction between hydrogen and oxygen is used to create a fuel cell.

_____ [1 mark]

iii Write a balanced symbol equation for the reaction between hydrogen and oxygen.

☐ ☐ → ☐ [1 mark]

b This is a diagram of a fuel cell.

i Explain what happens at the negative electrode.

_____ [1 mark]

ii Explain what happens at the positive electrode.

_____ [1 mark]

iii Explain what is meant by a **redox** reaction.

_____ [1 mark]

3 a There are advantages in using a fuel cell in a spacecraft. Write about **three** advantages.

_____
_____ [3 marks]

b The car industry is developing fuel cells. Explain **two** advantages.

_____
_____ [2 marks]

c Write about **one** problem that will need to be overcome by the car user when using a fuel cell to power a car.

_____ [1 mark]

d There are advantages in generating electricity using a hydrogen–oxygen fuel cell over conventional methods of generating electricity. Write about **two** advantages.

_____
_____ [2 marks]

C6 CHEMISTRY OUT THERE

108

Redox reactions

1 a Write down the word equation for the reaction of iron, water and air.

☐ ☐ ☐ → ☐   [2 marks]

b This is a redox reaction. Explain why. Use ideas about electrons in your answer.

_____
_____   [2 marks]

2 a Write down the **three** ways of preventing air and water reaching the surface of the iron.

_____
_____   [2 marks]

b i Iron can be protected by galvanising. Explain how.

_____   [1 mark]

ii Iron can be protected by sacrificial protection. Explain how.

_____   [1 mark]

iii Iron can be protected by tinning. Explain how.

_____   [1 mark]

3 a Four metals were put into solutions of metal salts. Look at the table of results.

	metal being added			
solution used	magnesium	zinc	iron	tin
magnesium sulfate	✗	✗	✗	✗
zinc sulfate	✓	✗	✗	✗
iron sulfate	✓	✓	✗	✗
tin sulfate	✓	✓	✓	✗

key:
✗ means nothing happens
✓ means the metal gets coated

b Zinc metal did not react with the solution of magnesium salt. Give a reason why.

_____   [1 mark]

c Write the word equation for a displacement reaction between zinc and a metal salt.

_____   [2 marks]

d Write a symbol equation for the reaction between magnesium and zinc sulfate $ZnSO_4$.

_____   [2 marks]

e Write an ionic equation for the reaction between zinc and an iron salt solution Fe^{++}.

_____   [2 marks]

4 a Why are redox reactions given the name 'redox'?

_____   [1 mark]

b In the reaction from iron metal, Fe, to an iron solution with Fe^{2+} ions, electrons are gained and lost. What happens?

	electrons lost or gained?	oxidised or reduced?
Fe		
Fe^{2+}		

[4 marks]

Alcohols

1 a Ethanol can be made by fermentation. Write down the word equation for the fermentation of glucose.

☐ → ☐ ☐ [2 marks]

b i There needs to be an optimum temperature chosen for fermentation. Explain why.

_____
_____ [2 marks]

ii There needs to be an absence of air during fermentation. Explain why.

_____ [1 mark]

c Ethanol often needs to be removed from the excess water. Describe how this is done.

_____ [1 mark]

d i The molecular formula of ethanol has two carbon atoms and one oxygen atom in it. It also has hydrogen atoms. Write down the molecular formula for ethanol.

_____ [1 mark]

ii Draw the displayed formula for ethanol.

[2 marks]

iii Pentanol is an alcohol with five carbon atoms. Draw the displayed formula for pentanol.

[2 marks]

iv What is the general formula for an alcohol with n carbon atoms.

_____ [1 mark]

2 a i Ethanol can also be made by the action of ethene and water. Which catalyst is used?

_____ [1 mark]

ii Which conditions are used in this reaction?

_____ [1 mark]

b i Write the word equation for this reaction.

_____ [1 mark]

ii Write the balanced symbol equation for this reaction.

_____ [2 marks]

3 Ethanol can be heated with a different catalyst to produce ethene.

a On the diagram label the catalyst used. [1 mark]

b Write the word equation for this reaction.

_____ [1 mark]

c Write the balanced symbol equation for this reaction. _____ [2 marks]

Chemistry of sodium chloride (NaCl)

1 a Sodium chloride can be mined as rock salt and used for gritting icy roads.
   Where are the UK mines?
   _____ [1 mark]

b What problem can occur as a result of mining salt?
   _____ [1 mark]

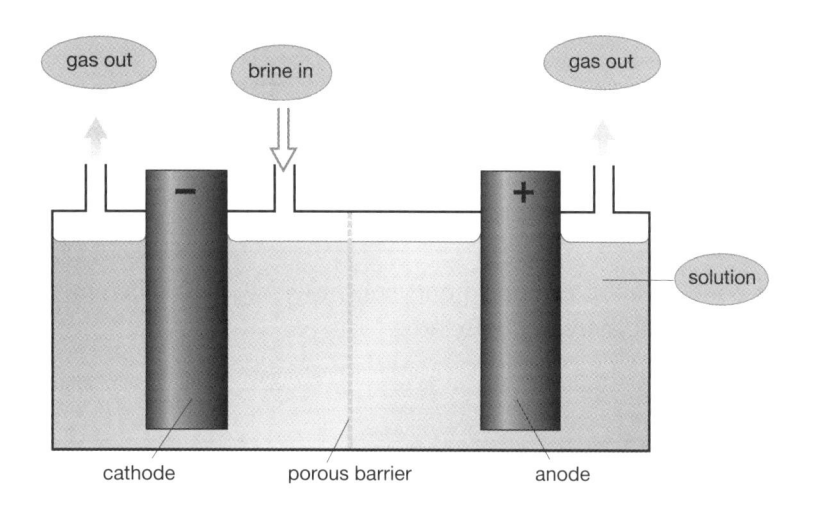

2 a Concentrated sodium chloride (brine) is electrolysed in this electrolysis cell.

i Which type of electrodes are used? _____ [1 mark]

ii At which electrode is chlorine given off? _____ [1 mark]

iii Another gas is given off at the other electrode. Which gas? _____ [1 mark]

iv Sodium hydroxide is also made during this process. Explain why.
   _____
   _____ [2 marks]

b i Write the equation for the reaction at the electrode where chlorine is given off.
   _____ [2 marks]

ii Write the equation for the reaction at the other electrode.
   _____ [2 marks]

c If **dilute** sodium chloride is electrolysed, chlorine is **not** given off.
   Which gas is given off? _____ [1 mark]

3 When molten sodium chloride is electrolysed there are two products at the electrodes.

a Write the equation for the reaction at the electrode at which sodium metal is deposited.
   _____ [2 marks]

b Write the equation for the reaction that happens at the other electrode.
   _____ [2 marks]

4 Which **two** chemicals are put together to make household bleach?
   _____ and _____ [2 marks]

Depletion of the ozone layer

1 CFCs been banned in the UK. Explain why.

_____

_____ [2 marks]

2 a A layer of ozone in the stratosphere can be depleted (reduced) by CFCs. When UV light acts on the CFC an atom is made from it that destroys the ozone.

Which atom? _____ [1 mark]

b This is a special type of atom. What is it called? Put a (ring) round the correct answer.

free electron   **free ion**   **free compound**   **free radical**

c Scientists are concerned that the effects of CFCs will be noticed for a long time.

Why will they be there for a long time?

_____ [1 mark]

d When a covalent bond breaks it can do so unevenly or evenly. Two different types of parts can form. Explain what these are.

1 _____

2 _____ [2 marks]

3 a What does the ozone layer let through if it is depleted?

_____ [1 mark]

b When CFCs react with ozone a chain reaction is set up. Write **two** symbol equations to show how the chain reaction keeps going.

_____

_____ [2 marks]

4 a Which kinds of compounds are now used as safer alternatives to CFCs?

_____ [1 mark]

b Look at the diagram of types of molecules that are sources of chlorine atoms in the stratosphere.

[Pie chart: natural 18%, mixed CFC 9%, synthetic CFC 51%, CCl_4/$C_2H_3Cl_3$ 22%]

What percentage of these gases are synthetic?

_____ [1 mark]

c CFCs will deplete ozone for a long time after they have been banned. Explain why.

_____

_____ [2 marks]

Hardness of water

1 a Rainwater dissolves carbon dioxide. How does this change the rainwater?
_____ [1 mark]

b Rainwater with carbon dioxide reacts with rocks made of calcium carbonate to form soluble calcium hydrogencarbonate. Write a word equation for this reaction.

☐ ☐ ☐ → ☐ [2 marks]

c The molecular formula for calcium hydrogen carbonate is $Ca(HCO_3)_2$.

Write a balanced symbol equation for the reaction of the $CaCO_3$ rocks with rain water.

☐ ☐ ☐ → ☐ [2 marks]

2 a Describe how an ion-exchange resin can soften water.
_____
_____ [2 marks]

b Explain why you have to put salt (NaCl) into a water softener after it has been used. Use ideas about ions in your answer.
_____
_____ [2 marks]

3 a Which chemical compound causes temporary hardness?
_____ [1 mark]

b Limescale ($CaCO_3$) is made when water with temporary hardness is boiled. It is removed by acid descalers. Write a symbol equation for the reaction of acid on limescale.
_____ [2 marks]

4 Look at the table.

sample of water	A	B	C	D
height of soap lather before boiling in mm	40	2	3	3
height of soap lather after boiling in mm	40	22	3	38

a One of the samples of water may have contained water that was permanently hard.

Explain how you know this from the data in the table.
_____
_____ [2 marks]

b Jo and Sam got these results when they carried out an investigation. They needed to make sure their experiments were fair tests. Write down **two** ways they did this.

1 _____
2 _____ [2 marks]

Natural fats and oils

1 a Oils and fats belong to one group of chemical compounds.

What is this group?

_____ [1 mark]

b Look at the two displayed formulae, M and N.

```
 H H H H H H H              H H H H    H H
 | | | | | | |              | | | |    | |
-C-C-C-C-C-C-C-           -C-C-C-C=C-C-C-
 | | | | | | |              | | |    | | |
 H H H H H H H              H H H    H H H
       M                          N
```

i M is a saturated compound. Explain how you know.

_____ [1 mark]

ii Circle the part of compound N that shows it is an unsaturated compound.

_____ [1 mark]

c Bromine water can be used to show the difference between M and N.

i Describe what you would do to test the difference.

_____ [1 mark]

ii Describe what you would **see**.

_____
_____
_____ [3 marks]

d You can test unsaturation with bromine water because of the changes you see.
Explain why those changes happen.

_____
_____ [1 mark]

2 a When soap is made, the chemical splits the oil making glycerol as well.

What is this process called?

_____ [1 mark]

b Write a word equation for the reaction between fat and a chemical to make soap and glycerol.

_____ [1 mark]

c What **type of reaction** is this process?

_____ [1 mark]

3 Suggest why unsaturated fats are healthier than saturated fats as part of our diet.

_____
_____
_____ [3 marks]

Analgesics

1 a Look at the displayed formula of one analgesic, X. The molecular formula is $C_8H_9O_2N$.

aspirin   X

Look at the displayed formula of aspirin. What is its molecular formula?

_____ [2 marks]

b i Look at the displayed formula of another analgesic Y.

Y

What part of their structures do aspirin, X and Y all have in common?

_____ [1 mark]

ii Look at analgesic X. Which **element** is in its structure that is not in the other two?

_____ [1 mark]

iii Look at aspirin and Y. They both have groups in the structure that are acid groups.

Find and circle the **two** acid groups –COOH. [2 marks]

c i There is an advantage in taking soluble aspirin. What is it?

_____ [1 mark]

ii Explain how the –COOH group is modified to make aspirin soluble.

_____
_____ [2 marks]

2 a Why is an analgesic, like aspirin, described as a drug?

_____
_____ [2 marks]

b Why must the chemicals used to make a drug be very pure?

_____ [1 mark]

c Explain how aspirin is manufactured from salicylic acid.

_____
_____
_____
_____ [4 marks]

C6 Revision checklist

- I know that a fuel cell is supplied with fuel and oxygen and the energy from the reaction between them is used to create a potential difference. ☐

- I know that the hydrogen–oxygen fuel cell uses the reaction: $2H_2 + O_2 \rightarrow 2H_2O$ ☐

- I know that the word equation for the displacement reaction between copper sulfate and zinc is: copper sulfate + zinc → zinc sulfate + copper. ☐

- I know that oxidation is the process of electron loss and reduction is electron gain. ☐

- I know that the word equation for the hydration of ethene is: ethene + water → ethanol. ☐

- I know that in the fermentation reaction, if the temperature is too high the enzyme in yeast is denatured. ☐

- I know that in the electrolysis of brine (concentrated sodium chloride solution), hydrogen is made at the cathode, chlorine is made at the anode and sodium hydroxide is also made. ☐

- I know that in the electrolysis of brine, the reaction at the cathode is: $2H^+ + 2e^- \rightarrow H_2$ and at the anode is: $2Cl^- - 2e^- \rightarrow Cl_2$ ☐

- I know that the formation of chlorine atoms (free radicals) in the stratosphere leads to the depletion of the ozone layer. ☐

- I know that $Cl\bullet + O_3 \rightarrow OCl\bullet + O_2$ and $OCl\bullet + O_3 \rightarrow Cl\bullet + 2O_2$ are part of the chain reaction in the stratosphere. ☐

- I know that boiling removes temporary hardness but not permanent hardness in water. Calcium hydrogencarbonate decomposes to give insoluble calcium carbonate. ☐

- I know that an ion-exchange resin will soften water by exchanging the Ca^{2+} ions in the water for Na^+ ions. The resin is recharged by running salt solution through the column. ☐

- I know that an unsaturated fat or oil has at least one carbon–carbon double bond. ☐

- I know that an unsaturated oil will change the colour of bromine water from brown to colourless. ☐

- I know that I can work out the molecular formula of aspirin from its displayed formula. ☐

- I know that the –COOH group on aspirin is reacted with a base to make a –COO⁻ ion, which makes the aspirin more soluble. ☐

Glossary

A
acidic A chemical that turns litmus paper red with a pH below 7.
alcohol A liquid produced when yeast respires sugar in the absence of oxygen.
alkali A substance which makes a solution that turns red litmus paper blue.
alkali metal A metal which burns to form a strongly alkaline oxide, e.g. potassium, sodium.
alkane Chemicals containing only hydrogen and carbon with the general formula C_nH_{2n+2}, e.g. methane.
alkene Chemicals containing only hydrogen and carbon with the general formula C_nH_{2n}, e.g. ethene.
allotrope Different forms of the same element made when the atoms of the element join together in different patterns, e.g. diamond and graphite are allotropes of carbon.
alloy A mixture of two or more metals.
ammonia An alkaline gas with the formula NH_3; it dissolves readily in water to make ammonium hydroxide solution.
analgesics A painkiller.
anion A negatively charged ion; it will move towards the anode in an electrolytic cell.
anode The positive electrode in a circuit or battery.
antioxidant Chemicals which delay the oxidation of other chemicals. They are important in paints, plastics and rubbers where they slow down degradation of the material. Vitamin C is an antioxidant in the body.
ascorbic acid The chemical name for vitamin C.
aspirin A painkilling drug originally made from the bark of willow trees.
atom The smallest part of an element, atoms consist of negatively charged electrons flying around a positively charged nucleus.
atomic number The number of protons in the nucleus of an atom.

B
base An alkali.
batch process A process that has a clear start and finish; often used to make medicines, at the end of the process a batch of product has been made and the reaction stops.
bauxite A mineral containing aluminium.
binding medium The substance, usually a thin glue, used to support pigments in paints.
biodegradable A substance which can be broken down by biological action in the environment.
biofuel A fuel that is produced from a living system, e.g. wood or ethanol produced from maize starch.
buckminster fullerene A very stable ball of 60 carbon atoms joined by covalent bonds; the whole structure looks like a geodesic dome.
bulk properties Properties that are independent of the amount of substance being measured, e.g. density or refractive index.
burette A glass device for measuring volumes of liquids very accurately; burettes are used in titrations.

C
carbohydrate Chemical found in all living things that contains the elements carbon, hydrogen and oxygen. Sugars are carbohydrates which dissolve in water and taste sweet. Starches are carbohydrates which cannot dissolve in water and do not taste sweet.
carbon A very important element, carbon is present in all living things and forms a huge range of compounds with other elements.
carbon dioxide A gas containing only carbon and oxygen; its chemical formula is CO_2.
carbon monoxide A poisonous gas containing only carbon and oxygen; its chemical formula is CO.
carbonate Compounds containing the carbonate group of atoms; the carbonate group formula is CO_3.
catalyst A chemical that speeds up a reaction but is not changed or used up by the reaction.
catalytic converter Boxes fitted to vehicle exhausts which reduce the level of nitrogen oxides and unburnt hydrocarbons in the exhaust fumes.
cathode The negative electrode in a circuit or a battery.
cation A positively charged ion; it moves towards the cathode in an electrolytic cell.
CFCs Chlorofluorocarbons are chemicals containing carbon, hydrogen, chlorine and fluorine; they were used in the past in refrigerators and as propellants in aerosols but have been phased out because they seem to damage the ozone layer.
chemical change A change that occurs when a number of substances react together to produce new substances.
chemical property The characteristic reactions of a substance.
chlorination Adding chlorine to a molecule or substance.
chromatography The science of producing chromatograms, chromatography can use paper or jelly-like films for the soluble substances to move along.
collision frequency The number of collisions in a particular area in a particular time.
colloid A mixture in which small particles of one substance are suspended in another.
combustion The reaction between a fuel and oxygen to form carbon dioxide and water; energy is released as light and heat.
compound A group of atoms bound together, in fixed proportions, by chemical bonds; compounds have different chemical and physical properties to the elements that they contain.
compression To push something together, to squeeze it and make it smaller.
concentration The amount of solute in a solution in g/dm^3.
conductor A substance that will let heat or electricity pass through it, e.g. copper.
Contact Process The industrial process for the manufacture of sulfuric acid using a catalyst (usually vanadium (V) oxide), temperatures of 450 °C and pressures of 200 kPa.
continental plate A large plate of solid rock in the Earth's crust containing a continental landmass.
continuous process A process which can continue indefinitely if new materials are added and wastes removed.

GLOSSARY

convection current Movement upwards of heated gases or liquids to float on top of the cooler, denser layers.

corrode Reaction of metals with the air to form powder or crystals, weakening the metal; the term rust is used when iron corrodes.

covalent bond A link between two atoms where electrons move around both atoms in the pair; covalent bonds tend to be strong bonds and form between non-metals.

cracking The breaking of large organic molecules into smaller ones using heat, pressure and sometimes catalysts.

critical temperature The temperature at which a key event takes place, perhaps a reaction starts or an animal dies.

crop yield The mass of useful material produced by a crop.

cryolite A compound of fluorine, aluminium and sodium (Na_3AlF_6) used to extract aluminium from bauxite by electrolysis.

crystals A solid substance with a regular shape made up of flat planes.

D

decompose To break apart.

delocalised electron An electron in a molecule that is not linked directly to an individual atom or covalent bond.

density The mass of an object divided by its volume.

diffusion Diffusion is the spreading of gases or liquids caused by the random movement of their molecules.

discharge To lose charge.

displacement reaction A reaction where one chemical, usually a metal, is forced out of a compound by another chemical, also usually a metal.

displayed formula A formula that shows all the bonds in the molecule as individual lines.

dissolve When a solid mixes in with a liquid so that it cannot be seen.

distillation Used to boil off a liquid from a mixture and then cool the vapours to produce a purer liquid.

dot and cross model A way to show how electrons are shared in covalent bonds.

ductile Can be drawn into a thin wire; metals are ductile.

E

efficient An efficient device transfers most of the input energy into the desired output energy.

electrode Bars of metal or carbon that carry electric current into a liquid.

electrolysis Using an electric current to split a compound – either in solution or in its molten state.

electrolyte The liquid carrying the electric current in an electric cell.

electromagnet A magnet produced by an electric current passing through a coil of wire.

electron A small negatively charged particle that orbits around the nucleus of an atom.

electron pattern The way electrons are arranged in a substance.

electronic structure The way electrons are arranged in the shells around an atom.

electrostatic To do with electric charges that are not moving; electrostatic charges behave differently to current electricity.

element A substance that cannot be split into anything simpler by chemical means, all the atoms of an element have the same atomic number although some may have different atomic masses.

emulsions Formed when tiny droplets of oil are dispersed through water (oil-in-water emulsion, e.g. milk) or water is dispersed through oil (water-in-oil emulsion, e.g. butter).

emulsifier A chemical which can help to break fats up into small globules so that they do not settle out of suspension.

endothermic A reaction that takes in energy when it happens.

equilibrium A balancing position.

eutrophication Waters with very high levels of minerals, often created by fertiliser pollution, producing a very heavy growth of algae.

exothermic A reaction which gives out energy when it happens.

F

fermentation Breakdown of food by microorganisms that does not require oxygen.

fertiliser A substance added to the ground by gardeners and farmers to help plants to grow.

filtration The process of separating large particles from small ones using a filter.

finite resource Resources that will run out because they are not being produced at the same rate as they are being used up.

formula A shorthand way to show the type and amount of elements present in a compound.

formula, empirical A formula that indicates the relative proportions of the elements in a molecule rather than the actual number of atoms of the elements.

fossil fuel A fuel formed by the decay of dead living things over millions of years, e.g. coal, oil and natural gas.

free radical A covalent bond, made of two electrons, when broken by UV, splits into equal halves to make two free radicals. Free radicals are highly reactive.

fuel Something that gives out energy, usually as light and heat, when it burns.

fuel cells Cells that use chemical reactions to generate electricity.

fullerene A cage-like arrangement of carbon atoms.

G

gas syringe A glass syringe used to measure volumes of gases very accurately.

giant ionic lattice A large collection of ions held together by strong electrostatic charges.

gradient A slope or difference in measurements between two areas, e.g. there is a concentration gradient between the water molecules inside and those outside a cell.

graphite A type of carbon often used in pencils as the 'lead'.

greenhouse gas Gases such as carbon dioxide and water vapour that increase the greenhouse effect.

H

Haber process The industrial process developed by Fritz Haber to make ammonia from nitrogen and hydrogen.

halogen A group of reactive non-metals with only one electron missing from their outer electron shell, e.g. chlorine and iodine.

hard water Water containing dissolved magnesium and calcium salts, mainly bicarbonates, which make it difficult for soap to form a lather.

hydrocarbon Hydrocarbon molecules are molecules that contain only carbon and hydrogen atoms. Many fuels are hydrocarbons, e.g. natural gas (methane) and petrol (a complex mixture).

hydrogen A colourless, odourless gas that burns easily in oxygen to form water; hydrogen is the lightest element.
hydrogen bond A force of attraction between the hydrogen atom in a molecule and a strongly electronegative atom such as nitrogen or oxygen.
hydrophilic A molecule or part of a molecule that dissolves easily in water; 'water loving'.
hydrophobic A molecule or part of a molecule that does not dissolve easily in water; 'water hating'.
hydroxide Chemicals containing an 'OH' group; hydroxides are often alkaline.

I

igneous Rocks formed from solidified molten magma.
indicator A chemical that changes colour in acid and alkaline solutions; indicators are used to find the pH of a solution.
insoluble A substance that will not dissolve; something that is insoluble in water may be soluble in other liquids.
intermolecular force A force between two molecules.
ion Charged particle made when an atom, or group of atoms, gains or loses electrons.
ionic exchange The exchange of ions to remove hardness in water using resins. Water flows over solid resin which traps calcium and magnesium ions on to it, taking these ions out of the water. They are exchanged for sodium ions.
ionic equation An equation showing the movement and behaviour of ions in a reaction.
isotope One of two or more atoms having the same atomic number but different mass numbers.

K

kinetic energy Energy due to movement.

L

lattice A regular arrangement of items, often applied to a collection of ions in a crystal.
lava Molten rock thrown up by a volcano.
limescale Deposit of calcium carbonate caused by boiling temporary hard water.
lithosphere The outer part of the earth, consisting of the crust and upper mantle, approximately 100 km thick.
lustrous Having a sheen or glow; highly polished metals such as gold are often described as lustrous.

M

magma Molten rock inside the Earth.
magnesium A lightweight, silvery white metal which burns with a very bright white flame.
malleable Can be beaten into flat sheets; metals are malleable.
mass Mass describes the amount of something - it is measured in kilograms.
mass number The mass of an atom compared with hydrogen.
melting point The temperature at which a solid changes to a liquid.
metal halide A compound containing only a metal and a halogen atom, e.g. sodium chloride.
metallic bond The bond typical of metals in which electrons are shared between many atoms in a stable crystalline structure.
metamorphic Rock formed when heat and pressure changes the characteristics of an existing rock.
methane A colourless, odourless gas that burns easily to give water and carbon dioxide.

mineral Natural solid materials with a fixed chemical composition and structure; rocks are made of collections of minerals; mineral nutrients in our diet are simple chemicals needed for health, e.g. calcium and iron.
molecular formula A formula that shows the number and kinds of atoms in a molecule.
molecule A group of atoms joined together by chemical links.
moles A mole of any substance contains 6×10^{23} particles of that substance, it weighs the atomic weight of the substance expressed in grams.
molten Something that has been heated to change it from a solid to a liquid.

N

nano properties The properties of materials at the nanoscale, often different to the same material's properties at the visible scale.
nanoparticle A particle that has at least one dimension that is smaller than 100 nanometres; a nanometre is 10^{-9} m.
nanoscale Objects and events occurring at distances of fewer than 100 nanometres.
nanotube A molecule consisting of carbon atoms joined in a cylinder one to two nanometres in diameter and about a millimetre in length.
naphtha A group of generally liquid chemicals derived from crude oil by distillation.
negative ion An ion with a negative charge.
neutral A neutral solution has a pH of 7 and is neither acid nor alkaline.
neutralise To react an acid with an alkali to produce a neutral solution.
neutron A particle found in the nucleus of an atom, it has no electrical charge and a mass of 1 atomic mass unit.
nitinol An alloy of nickel and titanium that has the ability to return to a predetermined shape when heated.
nitrogen A non-reactive gas that makes up most of the atmosphere.
nitrogenous Chemicals containing nitrogen.
non-biodegradable Objects that cannot be broken down by living organisms, e.g. many plastics, so they last for an extremely long time in the environment.
nucleus The control centre of the cell, the nucleus is surrounded by a membrane that separates it from the rest of the cell; the central part of an atom containing the protons and neutrons.

O

oils Fatty chemicals.
opaque Opaque objects will not let light through so they make a shadow, e.g. black paper.
optimum temperature The temperature range that produces the best reaction rate.
oxidation A reaction which adds oxygen to a compound or element, e.g. combustion and respiration.
oxidising Agent chemicals which supply oxygen or receive electrons in a chemical reaction, e.g. air and bleach.
oxygen A colourless gas with no smell that makes up about 20% of the air.
ozone layer A layer of the upper atmosphere that is particularly rich in the gas ozone.

P

particles A very small part of something.
patent A law that gives the creator of an invention the sole right to make, use or sell that invention for a given period of time.

Glossary

period A horizontal row of the periodic table.
periodic table A way of grouping elements according to their similarities, first devised by Dimitri Mendeleev.
permanent hardness Hardness in water due to calcium sulfate; it cannot be removed by boiling.
pH meter An electronic device for measuring the pH of a solution.
pH scale The range of levels of acidity or alkalinity; a pH of 7 is neutral, a pH below 7 is acid and the lower it goes the more acidic it becomes. A pH above 7 is alkaline.
phosphorescent Phosphorescent materials glow gently even after the original light source has been removed.
pigment Chemicals which absorb certain wavelengths of light and so look coloured.
pollutant A chemical that causes pollution.
polymer A molecule made of many repeating subunits, for example polythene or starch.
polymerisation The process of forming large polymers from smaller monomer molecules.
positive ion An ion with a positive charge.
precipitate A insoluble solid formed suddenly when two solutions react together.
product Something made by a chemical reaction.
protein A group of complex molecules that contain carbon, hydrogen, oxygen, nitrogen, and usually sulfur. They are made of one or more chains of amino acids. Proteins are important parts of all living cells and include enzymes, hormones, and antibodies. They are essential in the diet of animals for the growth and repair of tissue and can be obtained from foods such as meat, fish, eggs, milk, and legumes.
proton A particle found in the nucleus of an atom with a charge of plus one and a mass of one atomic mass unit.

R

reactant A chemical taking part in a chemical reaction.
reaction time The time taken for a reaction to finish.
reactivity A measure of how easily a chemical will react, usually applied to the reaction between metals and water.
recommended daily allowance The amount dieticians recommend for a healthy diet.
redox reactions Reactions that involve oxidation and reduction.
reducing agents Chemicals that take in oxygen or give electrons away in a chemical reaction, e.g. methane in a gas burner.
reduction The loss of oxygen or the gaining of hydrogen in a reaction.
relative atomic mass The mass of an atom or particle in comparison to the mass of hydrogen, which is taken as 1.
relative formula mass The mass of a molecule in comparison to the mass of hydrogen, which is taken as 1.
renewable Sources of energy that will not run out such as wind power, wave power and solar power.
resistance The amount by which a conductor prevents the flow of electric current.
respiration The chemical process that makes energy from food in the cells in your body. All living things must respire.
reversible reaction A reaction whose direction can be changed by a change in conditions.

S

salt A compound made when an acid reacts with an alkali.
saponification Heating vegetable oils with sodium hydroxide solution in large vats in the manufacture of soap.
saturated A solution that cannot dissolve any more solute.
sedimentary Rock formed when sediments from other rocks are laid down and compacted together.
sedimentation Particles settling out of suspension in water.
shell A grouping of electrons around an atom.
soluble A substance that can dissolve in a liquid, e.g. sugar is soluble in water.
solute Something that dissolves in a liquid to form a solution.
solution Something formed when a solute dissolves in a liquid.
solvent The liquid that dissolves a solute to make a solution.
stable electronic structure A configuration of electrons in the shells around an atom that produces a stable structure.
stable octet The most stable structure for the second shell around an atom.
strength of acid The degree to which an acid ionises.
superconductivity This occurs when the resistance of a substance to electricity falls to zero.
superconductor Something with zero electrical resistance.
surface area The area of a surface, which has a significant effect on the rate of many chemical reactions.
synthetic Made by humans, e.g. plastics are synthetic compounds which do not occur naturally.

T

tectonic plate Sections of the Earth's crust that float on top of the mantle. Plates are hundreds of miles across and move relative to each other by a few inches a year.
temporary hardness Hardness of water caused by calcium hydrogencarbonate. It can be removed by boiling the water.
tensile strength The maximum force a material can withstand before it snaps.
thermal To do with heat.
thermal decomposition Breaking down a chemical using heat.
thermochromic A pigment that changes colour when it gets hotter or colder.
titration Adding carefully measured amounts of a solution of known concentration to an unknown one to reach an end point which allows the concentration of materials in the unknown solution to be calculated.
transition element A metal belonging to the transition group in the periodic table.

U

Universal indicator solution An indicator that changes colour in solutions of different pH.
unsaturated A solution that can dissolve more solute. An unsaturated hydrocarbon can react with more hydrogen because it contains a number of double-carbon bonds.

Y

yeast A unicellular fungus used extensively in the brewing and baking industries.
yield The ratio of product to starting materials; a high yield means that most of the starting material is converted to useful products.

Answers

Here are the answers to the topic questions from the revision guide pages, and the answers to the exam-practice questions in the workbook.

This section is perforated, so that you can remove the answers to help test yourself or a friend.

C1 Carbon chemistry

Page 5
1 They change shape.
2 Irreversibly change shape.
3 Decomposes means that it breaks down. The three products are sodium carbonate, carbon dioxide and water.
4 $2NaHCO_3 \rightarrow Na_2CO_3 + H_2O + CO_2$

Page 6
1 Ascorbic acid (vitamin C).
2 It makes it more difficult for bacteria or mould to grow.
3 They are made up of two parts, a head and a tail. The tail is a 'fat-loving' part and the head is a 'water-loving' part.
4 The hydrophobic tail is attracted into the oil but the head is not. The hydrophilic head is attracted to water and 'pulls' the oil on the tail into the water.

Page 7
1 Acid and alcohol.
2 One person may object to cosmetics being tested on animals, as the animals may be harmed and they have no control over what happens to them. The other may say that they feel safer if the cosmetics have been tested on animals.
3 A substance that evaporates easily.
4 The force of attraction between a varnish molecule and a solvent molecule is stronger than between two varnish molecules. Water doesn't dissolve nail varnish because the force of attraction between two water molecules is stronger than that between a water molecule and a molecule of nail varnish.

Page 8
1 At the top.
2 Bitumen and heavy oil, because there are stronger forces of attraction between longer chains.
3 C_7H_{16}.
4 By cracking large hydrocarbons to make smaller more useful molecules.

Page 9
1 High pressure and a catalyst.
2
3 Alkanes do not have a double C=C bond, alkenes do.
4 C_nH_{2n}.

Page 10
1 It makes people sweat as it is not breathable.
2 A Gore-Tex® pore is 700 times larger than a water vapour molecule.
3 Toxic gases.
4 They have weak intermolecular forces of attraction between the polymer molecules, so the chains can slide over one another and can be stretched easily.

Page 11
1 Carbon dioxide is a greenhouse gas and contributes to climate change. It is a global problem that cannot be solved by one country and so needs careful agreement.
2 Limewater.
3 Less soot is made, more heat is released and toxic carbon monoxide gas is not produced.
4 $C_5H_{12} + 8O_2 \rightarrow 5CO_2 + 6H_2O$

Page 12
1 When energy is transferred to the surroundings in a chemical reaction (energy is released).
2 A reaction where there is more energy given out when the products are formed than the energy taken in to break the bonds of the reactants.
3 A yellow flame.
4 Energy transferred = $100 \times 4.2 \times 30 = 12600$
Energy per gram = $\frac{12600}{2} = 6300$ J/g

C2 Rocks and metals

Page 14
1 A colloid is formed when small solid particles are dispersed through the whole of a liquid, but are not dissolved in it.
2 First the solvent evaporates, then the binding medium is oxidised.
3 Emulsion, as the evaporating solvent is water not another type of solvent.

Page 15
1 Limestone and clay.
2 Under heat and pressure.
3 Calcium oxide and carbon dioxide.
4 $CaCO_3 \rightarrow CaO + CO_2$.

Page 16
1 They are less dense.
2 When two plates collide, the more dense oceanic plate sinks under the less dense continental plate. The rocks partially re-melt.
3 If the magma cools slowly, large crystals are made. Rapid cooling produces smaller crystals.
4 Silica-rich magma is less runny than iron-rich magma and produces volcanoes that may erupt explosively shooting out hot ash and pumice.

Page 17
1 The cathode.
2 Copper sulphate solution.
3 Copper and zinc.
4 They change shape.

Page 18
1 The oxide of aluminium becomes a protective layer.
2 Hydrated iron(III) oxide.
3 Aluminium is less dense and so the car is lighter.
4 Aluminium will corrode less or produce a lighter car than steel. Steel will be stronger and cost less to produce.

Page 19
1 Photosynthesis.
2 The process of photosynthesis is reduced so that carbon dioxide is not converted to oxygen.
3 Changes carbon monoxide to carbon dioxide.
4 $2CO + 2NO \rightarrow N_2 + 2CO_2$.

Page 20
1 For a reaction to take place, particles must collide often enough, with sufficient energy. If the particles move faster they will collide more successfully. If they are more crowded they will collide more often. In both cases the reaction will be faster.
2 As the concentration increases, the number of successful collisions per second increases and so the rate of reaction increases.
3 At a higher temperature the particles have more energy, so they collide more successfully, so the reaction is quicker. However, if the same mass of magnesium is used it will produce the same volume of hydrogen each time.
4 Extrapolation means extending a graph to read an expected reading that has not been measured.

Page 21
1 Any two from: sulfur, flour, custard powder, wood dust.
2 Half the mass of zinc will produce only half the volume of gas if the acid remains in excess both times.
3 There will be more collisions each second, which means the rate of reaction increases.
4 It helps reacting particles collide with the correct orientation and allows collisions between particles with less kinetic energy than normal to be successful.

C3 The periodic table

Page 23
1 1.
2 Chlorine-37.
3 11.
4 Magnesium.

Page 24
1 Electrons are gained.
2 A metal atom needs to lose electrons. The electrons transfer from the metal atom to a non-metal atom. A non-metal atom needs to gain electrons. The electrons transfer to the non-metal atom from the metal atom.
3
4 Because ions can move in the molten liquid.

Page 25
1
2 H_2O
3 Third.
4 The molecules are easy to separate.

Page 26
1 It is less dense than water.
2 Group 1 metals all have 1 electron in their outer shell.
3 Rubidium hydroxide.
4 The fourth shell is further away from the attractive 'pulling force' of the nucleus, so the electron from potassium is more easily lost than the electron from sodium. Potassium is therefore more reactive than sodium.

Page 27
1 The process of electron gain.
2 Potassium + bromine → potassium bromide.
3 $2Li + Br_2 \rightarrow 2LiBr$.
4 Iodine is less reactive than bromine so does not displace it.

Page 28
1 The formula of the compound breaking up is H_2O.
2 The H^+ ions migrate towards the cathode where they accept an electron. Two ions bond together to form a molecule of hydrogen gas, H_2.
3 They are gradually worn away by oxidation.
4 To lower the temperature of the melting point of aluminium oxide to save energy costs.

Page 29
1. Orange/brown.
2. In sodium hydroxide solution, Fe^{2+} ions form a grey/green solid and Fe^{3+} ions form an orange gelatinous solid.
3. copper carbonate → copper oxide + carbon dioxide.
4. $Cu^{2+} + 2OH^- → Cu(OH)_2$.

Page 30
1. It has a high thermal conductivity and is resistant to attack by metals or acids.
2. A metallic bond has a strong electrostatic force of attraction between close-packed positive metal ions and a 'sea' of delocalised electrons.
3. It levitates.
4. They only work at very low temperatures.

C4 Chemical economics
Page 32
1. Zinc oxide, zinc hydroxide or zinc carbonate.
2. Magnesium sulfate.
3. $ZnO + 2HCl → ZnCl_2 + 2H_2O$.
4. Hydroxide ions, OH^-.

Page 33
1. $40 + 2(14 + (3 × 16)) = 40 + 2(14 + 48) = 40 + 2(62) = 40 + 124 = 164$.
2. When chemicals react, the atoms of the reactants swap places to make new compounds, which are the products. They do not disappear.
3. 75%.
4. 22 g CO_2.

Page 34
1. 149.
2. The percentage that is nitrogen is $\frac{42}{149} × 100 = 28\%$.
3. Potassium hydroxide and phosphoric acid.
4. Titrate the potassium hydroxide with phosphoric acid, using an indicator. Repeat the titration until three consistent results are obtained. Use the titration result to add the correct amounts of acid and alkali together, without the indicator. Evaporate most of the water using a hot water bath. Leave the remaining solution to crystallise, then filter off the crystals.

Page 35
1. About 450°C.
2. The yield is lower than at a lower temperature but the rate of production is higher.
3. The yield would be too low.
4. Very high pressures are difficult and costly to maintain. A lower, optimum pressure is chosen to give a reasonable yield at a reasonable cost.

Page 36
1. An organic acid and an alkali.
2. The oil.
3. Fabrics may be damaged by washing in water and grease stains do not dissolve in water but do in dry-cleaning solvent.
4. Water to water.

Page 37
1. A batch process is where the whole process takes a limited time then stops and can be changed if necessary.
2. The plant can be used to full capacity at all times.
3. Any two from: there are legal requirements, investment costs of its research and development, raw materials, expensive extraction from plants, it is labour intensive.
4. Many years.

Page 38
1. Conducts electricity and has a high melting point.
2. Any two from: semiconductors in electrical circuits, industrial catalysts, reinforce graphite for tennis rackets.
3. Graphite has delocalised electrons that can move through the structure. Diamond does not.
4. To trap or transport molecules.

Page 39
1. Sand is used to filter out fine particles that do not sediment out.
2. The water needs to be heated up to boiling point and then cooled.
3. Precipitation.
4. $2AgNO_{3(aq)} + MgBr_{2(aq)} → 2AgBr_{(s)} + Mg(NO_3)_{2(aq)}$.

C5 How much?
Page 41
1. 90.
2. 1320 g.
3. C_3H_7.
4. H_2SO_4.

Page 42
1. Negative ions (cations).
2. $Pb^{2+} + 2e^- → Pb$ and $2I^- - 2e^- → I_2$.
3. Current and time.
4. 2160 C.

Page 43
1. A substance that dissolves in a liquid.
2. 15 g.
3. 90 cm^3.
4. 0.5.

Page 44
1. It increases.
2. Concentration $= \frac{\text{number of moles}}{\text{volume}}$.
3. To increase the reliability.
4. 0.113 mol/dm^3.

Page 45
1. 17 cm^3.
2. 0.04 moles.
3. Only half the mass of Mg was used in reaction 2, so only half the volume of gas is produced.
4. $14 ÷ 10 = 1.4$ cm^3/s.

Page 46
1. To the right.
2. Moves to the right.
3. 450 °C, atmospheric pressure, catalyst of vanadium pentoxide, V_2O_5.
4. High temperatures reduce the yield of the reaction (as it is an exothermic reaction) but the rate would be too low at low temperatures, so a compromise is found, 450 °C.

Page 47
1. Not all molecules separate into ions.
2. The ionisation of ethanoic acid in water is not complete. It is a reversible reaction where the position of equilibrium lies to the left.
3. H^+.
4. In strong acids the concentration of H^+ is higher, so collision frequency is greater.

Page 48
1. Silver nitrate + sodium iodide → silver iodide + sodium nitrate.
2. If the solution used to make the precipitate was potassium nitrate instead of sodium nitrate the same precipitate would occur as sodium, potassium and nitrate ions remain in solution.
3. To remove traces of the solution filtered with the precipitate.
4. $Ag^+_{(aq)} + Br^-_{(aq)} → AgBr_{(s)}$.

C6 Chemistry out there
Page 50
1. Petrol reserves will eventually run out. More laws are being passed to reduce the pollution from exhausts.
2. The main product from the reaction is water with no nitrogen oxides. Fuel cells cause fewer disposal problems than conventional batteries. Fuel cells weigh less, making cars lighter, reducing fuel consumption.
3. Heat is given out.
4. The reaction at the negative electrode. $2H_2 → 4H^+ + 4e^-$.

Page 51
1. Prevents oxygen or water reaching the surface of the iron.
2. Electrons are lost.
3. Zinc + iron nitrate → zinc nitrate + iron.
4. $Mg + FeSO_4 → MgSO_4 + Fe$.

Page 52
1. Glucose → ethanol + carbon dioxide.
2. Enzymes needed for the reaction will be denatured.
3. Hot aluminium oxide.
4. $C_2H_4 + H_2O → C_2H_5OH$.

Page 53
1. Using underground cutting machines. Pumping water down a borehole, dissolving the salt, pumping to surface and evaporating.
2. $2Na^+ + 2e^- → 2Na$.
3. Separates the products.
4. $2Cl^- - 2e^- → Cl_2$.

Page 54
1. They increase.
2. The ultraviolet part of the electromagnetic spectrum has exactly the right frequency to make ozone molecules vibrate. The energy of the ultraviolet is converted into movement energy inside each molecule.
3. Chlorine free radicals attack ozone molecules, turning the ozone back into oxygen gas and depleting the ozone layer.
4. $Cl• + O_3 → OCl• + O_2$.
   $OCl• + O_3 → Cl• + 2O_2$.

Page 55
1. Rainwater is acidic.
2. $MgCO_{3(s)} + CO_{2(g)} + H_2O_{(l)} → Mg(HCO_3)_{2(s)}$.
3. Both kinds of hardness.
4. To recharge the resin column with Na^+ ions.

Page 56
1. Bromine water is orange and is decolourised when shaken with an unsaturated compound/ it loses its colour.
2. People whose diet is rich in unsaturated oils usually have lower levels of cholesterol. High levels of cholesterol can lead to heart problems.
3. Vegetable oils are heated in large vats with sodium hydroxide solution.
4. When fats react with sodium hydroxide the molecules split into a molecule of glycerol and molecules of soap. The reaction is a hydrolysis reaction.

Page 57
1. Can cause liver damage.
2. The stomach lining is attacked causing slight bleeding in the stomach.
3. $C_{13}H_{18}O_2$.
4. The acid group –COOH is reacted with a base such as sodium hydroxide. The –COOH group forms the –COO$^-$ ion and Na+. This ionic part makes aspirin soluble in water.

Workbook answers

Remember: Check which grade you are working at.

Page 62 Fundamental concepts
1. a $2Mg + O_2 \rightarrow 2MgO$
   b $CuO + 2HNO_3 \rightarrow Cu(NO_3)_2 + H_2O$
   c $KCl + Pb(NO_3)_2 \rightarrow 2KNO_3 + PbCl_2$
2. a i 9 ii 1 Cu, 2 N, 6 O
   b i H_2SO_4 HCl HNO_3 CH_3COOH or $C_2H_4O_2$ ii NH_3 NaOH KOH
      iii NaCl $MgSO_4$ $CaCl_2$ K_2SO_4 iv $BaCl_2$ CO_2 H_2 H_2O
      v Na_2CO_3 C_2H_5OH or C_2H_6O $C_6H_{12}O_6$
3. a i Single covalent bond ii CH_3COOH or $C_2H_4O_2$
   b ionic c Shared pair of electrons
   d (diagram: methane + oxygen → carbon dioxide + water)

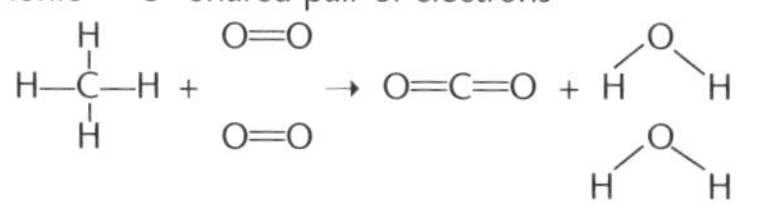

C1 Carbon chemistry
Page 63 Cooking
1. The texture of food is improved; the taste of food is improved; the flavour of food is enhanced; food is easier to digest *(Any 3 = 1 mark each)*
2. a Potatoes; flour b Meat; eggs
   c The protein molecules change shape; the shape change is irreversible; the protein molecule is said to be denatured; this changes the appearance/texture of the protein *(Any 3)*
3. a i Sodium hydrogencarbonate → sodium carbonate + carbon dioxide + water
      ii Sodium hydrogencarbonate
      iii Sodium carbonate; carbon dioxide; water
   b $2NaHCO_3 \rightarrow Na_2CO_3 + H_2O + CO_2$
4. Colourless; milky (cloudy)

Page 64 Food additives
1. a Stop food from reacting with oxygen and turning bad
   b Tinned fruit; wine c 56 J
2. a To stop food spoiling
   b Packaging that changes the condition of the food to extend its shelf life
   c It prevents the need for additives such as antioxidants to be added to foods
   d A catalyst
   e An indicator shows how fresh a food is on the outside of a package; a central circle darkens as the product loses its freshness
3. a The tail is a 'fat-loving' part and the head is a 'water-loving' part; the fat-loving part of the molecule goes into the oil and attracts it towards this end; the water-loving part will not go in; the water-loving part stays out of the oil but is attracted to the water molecules; the oil is, therefore, 'hooked up' to the water
   b The mayonnaise does not separate as the egg yolk has a molecule that has two parts; one part is a water-loving part that attracts vinegar to it, called the hydrophilic head; the other part is a water-hating part that attracts oil to it, called the hydrophobic tail; the hydrophobic tail is attracted into the lump of oil but the head is not; the hydrophilic head is attracted to water and 'pulls' the oil on the tail into the water

Page 65 Smells
1. a i Acid + alcohol → ester + water
      ii (Label to mixture in flask)
      iii (Label to upward condenser tube)
      iv At **X** the vapour is cooling down again and condensing back to a liquid
      v So that the mixture can be boiled/react for longer (without drying out)

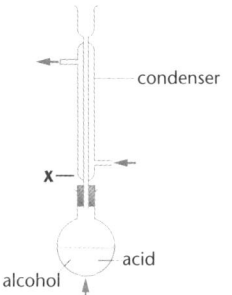

2. evaporates easily — it can be put directly on the skin
   non-toxic — its particles can reach the nose
   insoluble in water — it does not poison people
   does not irritate the skin — it cannot be washed off easily

3. a Solution b Solvents
4. a Particles of a liquid are weakly attracted to each other; when some particles of a liquid increase their kinetic energy the force of attraction between the particles is overcome and the particles escape through the surface of the liquid into the surroundings; this is evaporation; if this happens easily the liquid is said to be volatile
   b This is because the force of attraction between two water molecules is stronger than that between a water molecule and a molecule of nail varnish; also the force of attraction between two varnish molecules is stronger than between varnish molecule and water molecule

Page 66 Making crude oil useful
1. a A molecule containing carbon and hydrogen only
   b i (A: at the bottom, left-hand side, of the tower = 1 mark)
      ii (B: it 'exits' through the bottom of the tower = 1 mark)
      iii (C: at the top of the tower = 1 mark)
      iv Fractions with lower boiling points such as petrol/LPG

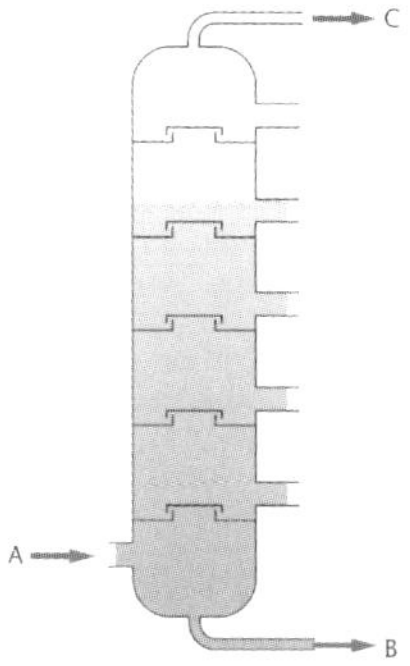

   c The forces between molecules are called intermolecular forces; these forces are broken during boiling/the molecules of a liquid separate from each other as molecules of gas; then either: the molecules in different fractions have different length chains; this means that the forces between the molecules are different; heavy molecules such as those that make up bitumen and heavy oil have very long chains; so they have strong forces of attraction between the molecules; this means that they are difficult to separate; a lot of energy is needed to pull each molecule away from another; they have high boiling points or: lighter molecules such as petrol have short chains; each molecule does not have very strong attractive forces and is easily separated; this means that less energy is needed to pull the molecules apart; they have very low boiling points *(Any 5 from either option)*

2. Oil slicks can harm animals, pollute beaches and destroy unique habitats for long periods of time; clean-up operations are extremely expensive and the detergents and barrages used cause problems
3. a C_7H_{16} b Alkenes have a double bond; alkanes have single bonds
   c Polymers
   d Cracking the fraction of heavy oil which is in excess supply to produce the smaller molecules needed for petrol which is in high demand but short supply

Page 67 Making polymers
1. a C b High pressure; catalyst c A double bond
   d (4 or 6 carbon atoms = 1 mark; alternate H and Cl atoms on bottom = 1 mark; brackets and bonds through either end = 1 mark; n at end = 1 mark) 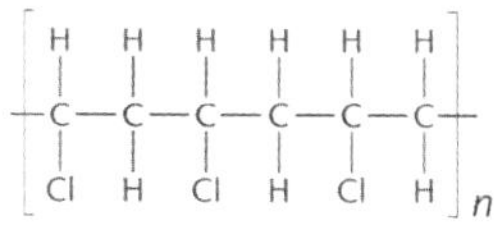
   e (Two carbon atoms joined by a double bond = 1 mark; CH_3 on top right hand side = 1 mark; only 4 other atoms/groups joined to two carbon atoms = 1 mark) 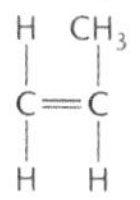
   f The reaction needs high pressure and catalyst; this causes the double bond in the monomer to break; and each of the two carbon atoms forms a new bond; the reaction continues until it is stopped, making a long chain
2. a i It has an oxygen atom in its structure
      ii It contains a double bond
      iii A polymer made from the monomer butene
   b i This is because the bromine solution has reacted with the alkene and has formed a new compound
      ii Remains orange

Page 68 Designer polymers
1. a White dental filling is better than a mercury amalgam; waterproof plastics are better than fabric plasters
   b i Hydrophobic means water-hating; the material repels water
      ii Water vapour from sweat can pass through the membrane but rainwater cannot so it keeps people dry when sweating; the membrane has pores which are 700 times larger than a water vapour molecule and therefore moisture from sweat passes through
2. a So that they do not have to be disposed of in landfill sites or burned but can decay by bacterial action
   b To make laundry bags for hospital so that they degrade when washed leaving the laundry in the machine *(Or any other suitable use)*
   c Landfill sites; waste valuable land; burning waste plastic: toxic gases; recycling: difficulty in sorting different polymers
3. a i (See diagram) ii (See diagram) 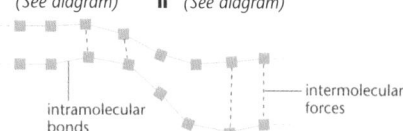
   b Some plastics have weak intermolecular forces of attraction between the polymer molecules so the polymer molecules can slide over one another/separate easily; some other plastics form intermolecular chemical bonds or cross-linking bridges between polymer molecules; these are strong so the polymer molecules cannot slide over one another; they are rigid/the chains cannot easily separate

Page 69 Using carbon fuels
1. a i Coal
      ii High energy value; good availability
   b Petrol and diesel are liquids so they can circulate easily in the engine; they are also stored easily in petrol stations along road networks; as these fuels are so easy to use and the population is increasing; more fossil fuels are being consumed, resulting in more carbon dioxide; this is a greenhouse gas; contributes to climate change which is a global problem *(Any 4)*
2. a Hydrocarbon fuel + oxygen → carbon dioxide + water
   b i Carbon dioxide
      ii Water (steam) is tested by turning white copper sulfate to blue
   c Less soot is made; more heat is released; toxic carbon monoxide gas is not produced
   d People who live in the house are in danger of being made ill or even dying from carbon monoxide poisoning if the room is not well ventilated/heater faulty
   e $C_3H_8 + 5O_2 \rightarrow 3CO_2 + 4H_2O$ *(Correct product formulae = 1 mark, correct balancing = 1 mark)*

Page 70 Energy
1. a Exothermic; endothermic; exothermic; exothermic
   b Endothermic
   c Bonds are broken which is an endothermic reaction; new bonds are made which is an exothermic reaction; as less energy is needed to break bonds than make new bonds then a reaction is exothermic overall
2. Blue; complete; yellow; incomplete

123

3 a Measure the same mass of water in two beakers; put the burners under the beakers for the same time with the same rate of gas; measure the gas volume with a meter; measure the temperature of the water before and after the experiment/temperature increase; same mass water; same volume gas
  b i Same distance of the calorimeter from the flame; repeating the experiment 3 times and excluding draughts
  ii The energy transferred is calculated using the formula:
  energy transferred = mass of water × 4.2 × temperature change
  energy = 100 × 4.2 × 50
  = 420 × 50
  = 21000 J

  energy per gram = $\frac{\text{energy supplied}}{\text{mass of fuel burnt}}$
  = $\frac{21000}{4.00}$
  = 5250 J/g

C2 Rocks and metals
Page 72 Paints and pigments
1 a When emulsion paint has been painted onto a surface as a thin layer, the water evaporates leaving the binding medium and pigment behind; as it dries it joins together to make a continuous film
  b Oil paint and emulsion paints are colloids because they are a mixture of solid particles in a liquid; the particles/droplets are very small and stay scattered in the liquid; they do not settle out in the bottom; particles in paint are small enough to stay dispersed through the liquid while it is in use
  c The solvent evaporates the oil slowly; reacts with oxygen in the air; the oil binding medium is oxidised by the air to form a tough, flexible film over the wood
2 a Used to paint cups; used to paint kettles; act as a warning (Any 2)
  b The thermochromic paints are mixed with different colours of normal acrylic paints in the same way that any coloured paints are mixed
  c The mixture contains a blue thermochromic pigment and yellow acrylic paint; when the mixture gets hot the thermochromic pigment goes colourless; the green loses the blue colour so only the yellow of the acrylic paint is seen
  d Phosphorescence
  e Light
3 a In luminous clock dials
  b Radioactive paints because the phosphorescent pigments are much safer

Page 73 Construction materials
1 a Limestone; marble; granite
  b

building material	brick	cement	glass	iron	aluminium
raw material	clay	limestone and clay	sand	iron ore	aluminium

  c Sedimentary
  d Granite is an igneous rock made from molten rock which solidifies slowly forming interlocking crystals; marble is a metamorphic rock made by putting chalk/limestone under heat and pressure
  e Limestone is made from the shells of dead sea-creatures that were compressed together
2 a Calcium carbonate → calcium oxide + carbon dioxide (Each compound = 1 mark)
  b Limestone is heated; clay is added
  c $CaCO_3 \rightarrow CaO + CO_2$
  d If a heavy load is put on a concrete beam it will bend very slightly; when a beam bends its underside starts to stretch; which puts it under tension and cracks start to form; steel is strong under tension; steel rods reinforce concrete to stop it stretching

Page 74 Does the Earth move?
1 a Less dense   **b** Continental plates; oceanic plates
  c In plate tectonics, energy from the hot core is transferred to the surface by slow convection currents in the mantle; if two adjacent convection currents move clockwise/anticlockwise the plates will be pulled together; if anticlockwise/clockwise the plates will move apart; or they can scrape sideways past each other 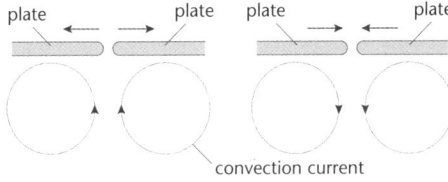
  d Two convection currents moving up and away from each other cause the plates to move apart; two convection currents moving down and towards each other cause the plates to collide; they can scrape sideways past each other
  e The coastlines of Africa and South America 'fit' suggesting that they have split and the continents have drifted apart; the sea floor shows ridges in the middle of the ocean suggesting plate movement
2 a Because it is less dense than the crust
  b

small crystals	large crystals
cool rapidly	cool slowly
basalt	granite

  c i Silica-rich magma
  ii It shoots out as clouds of searingly hot ash and pumice; the falling ash often includes large lumps of rock called volcanic bombs; geologists investigate past eruptions by looking at the ash layers; in each eruption coarse ash falls first, followed by fine ash, producing graded bedding (Any 2)

Page 75 Metals and alloys
1 a It must first be analysed to find out how much of each element is present
  b It has to be electrolysed again before it can be used
2 a Impure copper   **b** It is 'plated' with new copper
  c It dissolves
  d The cathode is plated in pure copper and gets thicker
  e Blister copper/boulder copper   **f** They sink to the bottom of the cell
3 a amalgam — contains mercury
  solder — contains lead and tin
  brass — contains copper and zinc
  b Pure copper conducts electricity so well
  c They can be bent more than steel so they are much harder to damage; they change shape at different temperatures, this is 'shape memory'
  d In the frames of glasses to stop them breaking; in shower heads to reduce the water supply if the temperature gets so hot it scalds; a small piece of metal can be put into a person's blocked artery and then warmed slightly, as it warms up it changes shape into a much larger tube that holds the artery open and reduces the risk of a heart attack (Any 2)
  e Nickel and titanium

Page 76 Cars for scrap
1 a Salt accelerates rusting which means that car bodies rust quicker
  b It has a protective layer of aluminium oxide which does not flake off the surface   **c** It flakes off
  d iron + water + oxygen → hydrated iron(III) oxide
2 a Iron; carbon
  b Stronger, harder; does not rust as easily as pure iron (Any 2)
  c Advantages: the mass of a car body made of aluminium will be less than the same car body made from steel; the car body made of aluminium will corrode less; disadvantage: the car body of the same car will be more expensive if made from aluminium
  d

material and its use	reasons material is used
aluminium in car bodies and wheel hubs	less dense so greater fuel economy, corrodes less
copper in electrical wires	good electrical conductivity
plastic in dashboards, dials bumpers	easily cleaned, lightweight, not easily damaged
pvc in metal wire coverings	flexible, easy to colour
plastic/glass composite in windscreens	transparent, less easy to break
fibre in seats	soft, comfortable, flexible

3 More recycling of metals means that less metal ore needs to be mined; recycling of iron and aluminium saves money and energy compared to making iron from their ores; less crude oil is used to make plastics; less non-biodegradable waste from plastics is dumped; recycling batteries reduces the dumping of toxic materials into the environment (Any 3)

Page 77 Clean air
1 a (See diagram)
  b (See diagram)
  c Combustion and respiration increase the level of carbon dioxide and decrease the level of oxygen; photosynthesis decreases the level of carbon dioxide and increases the level of oxygen

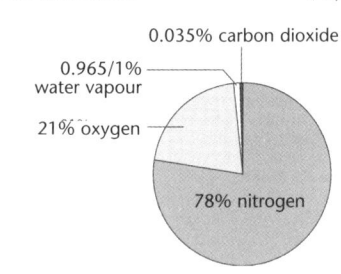

0.035% carbon dioxide
0.965/1% water vapour
21% oxygen
78% nitrogen

2 a

increased energy usage	more fossil fuels are being burnt in power stations
increased population	the world's energy requirements increase
deforestation	as more rainforests are cut down less photosynthesis takes place

  b i The original atmosphere contained ammonia and later carbon dioxide
  ii A chemical reaction between ammonia and rocks produced nitrogen and water; the percentage of nitrogen slowly increased; nitrogen is very unreactive so very little nitrogen was removed
  iii Much later organisms that could photosynthesise evolved; these organisms converted carbon dioxide and water into oxygen; as the percentage of oxygen in the atmosphere increased, the percentage of carbon dioxide decreased, until today's levels were reached

3 a

pollutant	carbon monoxide	oxides of nitrogen	sulfur dioxide
origin of pollutant	incomplete combustion of petrol or diesel in car engine	formed in the internal combustion engine	formed when sulfur impurities in fossil fuels burn

  b A rhodium catalyst
  c i The two gases formed are natural components of air/no longer pollutants
  ii Carbon monoxide + nitric oxide → nitrogen dioxide + carbon dioxide (Reactants = 1 mark, product = 1 mark)
  iii $2CO + 2NO \rightarrow N_2 + 2CO_2$ (Reactants = 1 mark, products = 1 mark, balance = 1 mark)

Page 78 Faster or slower (1)
1 a i *(See graph)* ii The higher temperature
      iii As the temperature increases the particles move faster; the reacting particles have more kinetic energy and so the number of collisions increases and the number of successful collisions increases

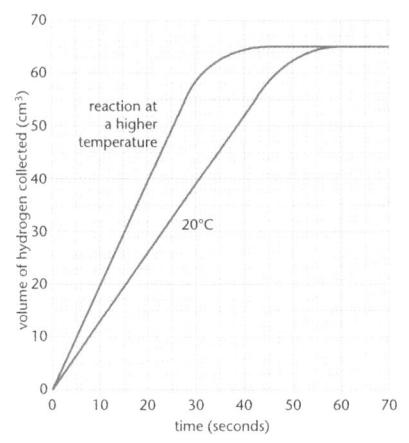

   b The frequency of successful collisions
   c It increases by the number of successful collisions between reactant particles that happen each second
2 a i The higher concentration of acid results in a quicker reaction rate
      ii The same amount of hydrogen collected in both cases
   b *(See graph)*

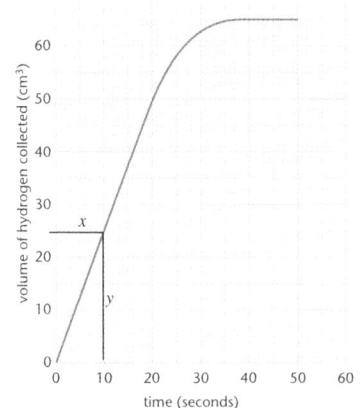

      i $x = 10$ sec; $y = 25$ cm^3
         Gradient $= \frac{y}{x} = 2.5$ cm^3/sec
      ii Reading within the graph between closer points

Page 79 Faster or slower (2)
1 Carbon dioxide and water vapour; a large volume of gaseous products are released, moving outwards from the reaction at great speed, causing the explosive effect

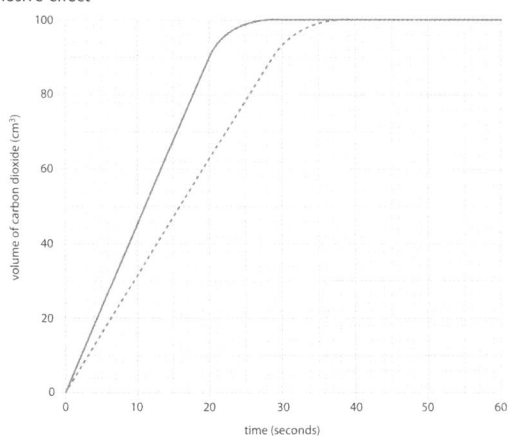

2 a A gas is given off: carbon dioxide
   b i 27–29 seconds   ii *(See graph)*
      iii As the reactants are used up there are less frequent collisions
   c There are more frequent collisions as there are more exposed particles
3 a It makes the reaction go faster
   b It often provides a specific surface on which the appropriate chemicals can react

C3 The periodic table
Page 81 What are atoms like?
1 a Protons; neutrons   *(Both = 1 mark)*
   b

	relative charge	relative mass
electron	–1	0.0005 (zero)
proton	+1	1
neutron	0	1

   c The number of protons in an atom
   d The total number of protons and neutrons in an atom
   e Fluorine
   f i 17
      ii 18
2 a Isotopes are elements that have the same atomic number but different mass numbers
   b

isotope	electrons	protons	neutrons
$^{12}_{6}C$	6	6	6
$^{14}_{6}C$	6	6	8

3 The first shell can only take 2 electrons; the second shell can only take up to eight; which is why a 3rd shell is needed

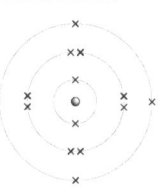

Aluminium

Page 82 Ionic bonding
1 a An atom which has extra electrons in its outer shell and needs to lose them to be stable
   b 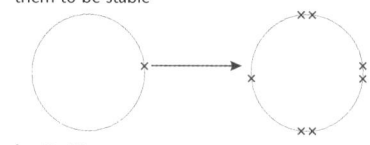
   c i Positive
      ii Magnesium/calcium
   d i Gaining
      ii Fluorine/chlorine/bromine/iodine
   e Positive; negative; lattice
   f
   g
   h Sodium chloride solution; molten (melted) magnesium oxide; molten sodium chloride
   i i High
      ii Conduct electricity
      iii Ions can move

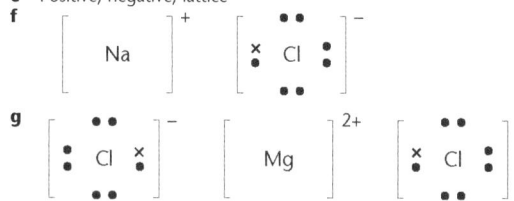

Page 83 Covalent bonding
1 a Covalent bonding
   b A molecule of water is made up of three atoms, two hydrogen and one oxygen; oxygen has six electrons in its outer shell, it needs two more electrons to be complete; hydrogen atoms each have one electron in their only shell; the oxygen outer shell is shared with each of the hydrogen electrons, so each of the hydrogen atoms has a share of two more electrons making the shell full
   c Because they are covalently bonded
   d i H$_2$O   ii CO$_2$

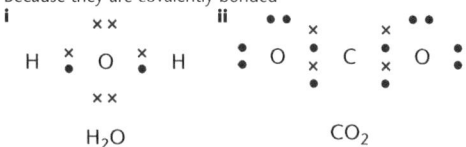

2 They are simple molecules with weak intermolecular forces; they are easy to separate so the substances have low melting points
3 a The group number is the same as the number of electrons in the outer shell; it has 1 electron in the outer shell
   b 7
   c i 2
      ii There are electrons occupying 2 shells
   d 6

Page 84 The group 1 elements
1 a i Hydrogen gas
      ii Their density is less than the density of water
   b Sodium + water → sodium hydroxide + hydrogen
   c

reactivity increases down	melting point in °C	boiling point in °C
$_3$Li	179	1317
$_{11}$Na	98	892
$_{19}$K	64	774

   d Group 1 metals all have 1 electron in their outer shell so they react in a similar way
   e 2Na + 2H$_2$O → 2NaOH + H$_2$
2 a They moistened a flame test wire with dilute hydrochloric acid; dipped the flame test wire into the sample of solid chemical; held the flame test wire in a blue Bunsen burner flame; put on safety goggles; recorded the colour of the flame in a table
   b red — potassium; yellow — sodium; lilac — lithium (red→potassium, yellow→sodium, lilac→lithium as shown by crossed lines: red–potassium, yellow–lithium shown crossed, lilac–sodium)

red → potassium
yellow → lithium
lilac → sodium

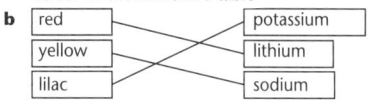

3 a Li − e⁻ → Li⁺/Li → Li⁺ + e⁻
  b Lithium loses its outer electron from its second shell; sodium loses its outer electron from its third shell; the third shell is further away from the attractive 'pulling force' of the nucleus so the electron from sodium is more easily lost than the electron from lithium; sodium is therefore more reactive than lithium
  c Oxidation

Page 85 The group 7 elements
1 a
| chlorine | green gas |
| iodine | grey solid |

  b They all have seven electrons in their outer shell
  c i Cl + e⁻ → Cl⁻
     ii Reduction
     iii Fluorine gains its last electron into its second shell; chlorine gains its last electron into its third shell; the third shell is further away from the attractive 'pulling force' of the nucleus so the electron to fluorine is more easily gained than the electron to chlorine; fluorine is therefore more reactive than chlorine
2 a Potassium + iodine → potassium iodide
  b 2K + I₂ → 2KI
3 a i Chlorine displaces the bromide ions which become bromine solution which is red-brown/ a displacement reaction occurs
     ii This is because chlorine is more reactive than bromine so bromine does not displace the chloride ions
  b i Bromine + potassium iodide → potassium bromide + iodine
     ii Br₂ + 2KI → 2KBr + I₂

Page 86 Electrolysis
1 a The electrolyte is a dilute solution of sulfuric acid; two electrodes are connected to a dc source of electric current, between 6 V and 12 V, and placed into the electrolyte; the electrode connected to the negative terminal is the cathode; the electrode connected to the positive terminal is the anode; when the current is switched on bubbles of gas appear at both electrodes; water splits into two ions: H+ is the positive ion and OH⁻ is the negative ion; H+ is attracted to the negative cathode and discharged as hydrogen gas, H₂; OH⁻ is attracted to the positive anode and discharged as oxygen gas, O₂
  b Because the formula of the compound breaking up is H₂O
2 a At the cathode: the water and sodium chloride split up into ions and the ions are free to move; the positive H+ and Na+ ions migrate towards the negative cathode; only the H+ ions are discharged; each H+ ion gains one extra electron from the cathode; a pair of atoms then bonds to become a molecule of hydrogen that forms part of the gas; 2H+ + 2e⁻ → H₂ *(Any 3)*
  b The negative hydroxyl OH⁻ and Cl⁻ ions migrate to the positive anode; only the OH⁻ ions are discharged; four OH⁻ ions each gain one electron and combine to form an oxygen molecule and two water molecules; 4OH⁻ − 4e⁻ → 2H₂O + O₂ *(Any 3)*
3 a The ore of aluminium oxide is bauxite; aluminium oxide is melted; aluminium is formed at the graphite cathode; oxygen is formed at the graphite anode; the anodes are gradually worn away by oxidation; this forms carbon dioxide; the process requires a high electrical energy input
  b Aluminium oxide → aluminium + oxygen
  c At the cathode: electrons have been gained; this is an example of reduction; Al³⁺ + 3e⁻ → 3Al
     at the anode: electrons have been lost; this is an example of oxidation; 2O²⁻ − 2e⁻ → 2O₂
  d Aluminium oxide requires large amounts of electricity to melt at very high temperatures, which is very expensive; cryolite lowers the melting point

Page 87 Transition elements
1 a i Copper compounds are blue
     ii Iron(II) compounds are pale green
     iii Iron(III) compounds are orange/brown
  b i Iron is used in the Haber process to make ammonia
     ii It is a transition metal, because it is in the transition metal block
  c Copper carbonate → copper oxide and carbon dioxide
2 a
ion	colour
Cu²⁺	form a blue gelatinous solid
Fe²⁺	form a grey/green gelatinous solid
Fe³⁺	form an orange gelatinous solid

3 a CuCO₃ → CuO + CO₂
  b Fe³⁺ + 3OH⁻ → Fe(OH)₃

Page 88 Metal structure and properties
1 a Lustrous; malleable
  b Resistant to attack by oxygen/acids; good thermal conductivity; malleable
  c It has a low density
  d *(It is close-packed positive metal ions; held together by a strong electrostatic force of attraction; between a 'sea' of delocalised electrons)*

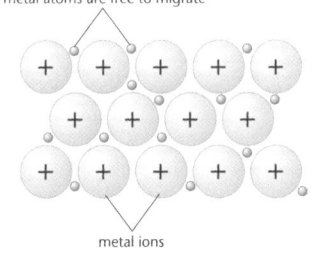

  e Because a lot of energy is needed to overcome the strong attraction between the delocalised electrons and the positive metal ions
2 a Materials that conduct electricity with little/no resistance
  b Loss-free power transmission; super-fast electronic circuits; powerful electromagnets *(Any 2)*
  c Superconductors only work at very low temperatures
  d Because delocalised electrons within its structure can move easily
  e i C
     ii A

C4 Chemical economics
Page 90 Acids and bases
1 a i An alkali is a base which dissolves in water
     ii Acid; water
  b Copper carbonate + sulfuric acid → copper sulfate + water + carbon dioxide
  c The salt formed is sodium nitrate
  d i NaCl   ii 2HCl + CaCO₃ → CaCl₂ + H₂O + CO₂
2 a Hydrogen ions, H⁺
  b Hydroxide ions OH⁻
  c Water H₂O
3 a The pH at the start is low as alkali is added the pH increases
  b The colour starts as purple. The pH falls as the acid neutralises the alkali; colour changes to blue; when neutral, the pH = 7; Colour is green

Page 91 Reacting masses
1 a 23 + 16 + 1 = 40
  b 40 + 12 + (16 × 3) = 100
  c 40 + 2 (16 + 1) = 40 + (2 × 17) = 74
  d When chemicals react, the atoms of the reactants 'swap' places to make new compounds – the products; these products are made from just the same atoms as before; there is the same number of atoms at the end as there were at the start, so the overall mass stays the same
2 a i 28
     ii 42
     iii $\frac{\text{Actual yield}}{\text{predicted yield}} \times 100$ = percentage yield
     iv $\frac{28 \times 100}{42} = 66\%$
  b i ZnCO₃ → ZnO + CO₂
     ii Relative formula mass of ZnCO₃ is 65 + 12 + 16 + 16 + 16 = 125; relative formula mass of CO₂ is 12 + 16 + 16 = 44; so if 125 g ZnCO₃ gives 44 g CO₂; then 12.5 g ZnCO₃ gives $\frac{12.5}{125} \times 44$
        = 4.4 g CO₂

Page 92 Fertilisers and crop yield
1 a To increase their crop yields
  b They are dissolved in water so they can be absorbed by plants through their roots
  c It is needed to make plant protein for growth.
2 a (NH₄)₂SO₄
     Mr = 2(14 + 4) + 32 + (16 × 4) = 132
  b The relative formula mass of KNO₃ is 39 + 14 + (16 × 3) = 101; mass of nitrogen = 14;
     the percentage that is nitrogen is $\frac{14}{101} \times 100 = 13.8\%$
3 a Water
  b i Phosphoric acid
     ii Ammonium hydroxide
     iii Phosphoric acid + ammonium hydroxide → ammonium phosphate + water
  c Phosphoric acid is reacted with ammonium hydroxide; the amounts used in the reaction must be exactly right, so a titration is carried out; titrate the alkali with the acid, using an indicator; repeat the titration until three consistent results are obtained; this is a neutral solution of potassium nitrate fertiliser, but it is contaminated with indicator; use the titration result to add the correct amounts of acid and alkali together without the indicator; the fertiliser made is dissolved in water, so most of the water is evaporated off using a hot water bath; leave the remaining solution to crystallise, then filter off the crystals
4 Eutrophication is caused by fertiliser run off increasing the amount of algae; this algal bloom prevents sunlight from reaching the plants below the water; anaerobic bacteria use up the oxygen in the water so fish die

Page 93 The Haber process
1 a Nitrogen is obtained from the air; hydrogen comes from natural gas; the gases are passed over an iron catalyst; under high pressure; an optimum temperature of 450 °C is chosen; there is a recycling system for unreacted nitrogen and hydrogen *(Any 3)*
  b i A higher pressure increases the percentage yield but high pressures costs more
     ii The high temperature decreases the percentage yield; however, higher temperatures make the reaction go faster
     iii 450 °C is an optimum temperature; the yield is not as good, but that yield is made faster, so a satisfactory amount is produced in the right time
2 a 400 atmospheres
  b Increases
  c Decreases
  d High temperature means higher rates but lower yields so it runs at the optimum temperature of 450 °C; this temperature means higher energy costs and also lower yields, but the increase in rate compensates; the plant produces more ammonia in a day at this temperature than it would at lower temperatures; total energy costs are not only due to heating costs; the plant needs compressors and pumps to achieve a high pressure; high pressure costs more so a lower, optimum pressure is used; although the reaction has a low percentage yield, the unreacted chemicals are recycled, and can go back into the reaction vessel, saving costs *(Any 3)*

Page 94 Detergents
1 a Organic acid + alkali → detergent (salt) + water
  b It is suitable for cleaning uses because; it dissolves grease stains; it dissolves in water at the same time *(Any 2)*
  c i It is better to wash clothes at 40 °C instead of at high temperatures because washing machines have to heat up a lot of water; this needs energy; the lower the temperature of the water the less energy is used and less greenhouse gases are put into the atmosphere
     ii As many dyes are easily damaged by high temperatures; it also means that many more fabrics can be machine washed as their structure would be damaged at higher temperatures
  d The hydrophobic part of the molecule goes into the grease; 'hydrophobic' means water-fearing; the molecule forms bonds with the oil or grease; the hydrophilic part of the molecule forms bonds with the water; 'hydrophilic' means water-loving; the molecule forms bonds with the water and 'pulls' the grease off the fabric/dish into the water; the grease is pulled off the surface by the hydrophilic part forming bonds with water and lifting it away

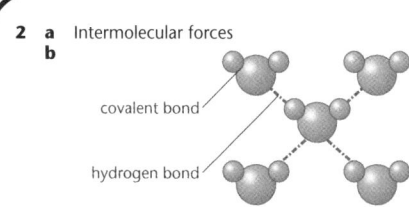

2 a Intermolecular forces
b [diagram showing covalent bond and hydrogen bond between water molecules]

Page 95 Batch or continuous?
1 a New batches are made when the stored medicine runs low; if a lot of one medicine is needed, several batches can be made at the same time; once they have made a batch of one drug it is easy to switch to making a different drug *(Any 1)*
b The process keeps going all the time so can be automated more easily
c Because it works at full capacity all the time it costs an enormous amount to build, but once running, it makes a large amount of product; employs very few people; making the cost per tonne very small
d They are flexible; it is easy to change from making one compound to another

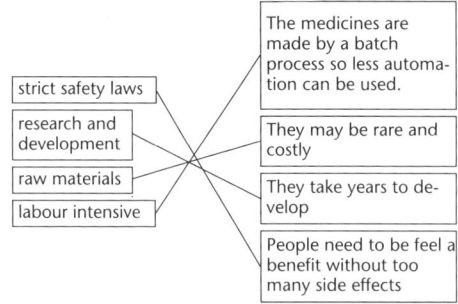

2 [matching diagram]
- strict safety laws — The medicines are made by a batch process so less automation can be used.
- research and development — They may be rare and costly
- raw materials — They take years to develop
- labour intensive — People need to be feel a benefit without too many side effects

3 There has to be an anticipated demand for the drug and a potential market for it; there will be huge research and development costs as the time taken is often up to ten years; it is a very expensive process as labour costs are high; promising compounds often have dangerous side-effects so lots of similar compounds have to be made to find the best one with the fewest side-effects; tests can be made on as many as 10,000 compounds to find one effective drug; a new drug must be trial tested for safety, eventually using human trials and then submitted for approval; only then can it start earning money; the company patents the drug, so for the next 20 years or so they can sell it for a high price; the drugs company has to recoup all its development costs in that time as they have to cover the initial investment over the pay-back period; after that other companies can make their own version of the drug; these are called generic drugs *(Any 6)*

Page 96 Nanochemistry
1 a [matching diagram: diamond, graphite, buckminster fullerene to structural diagrams]

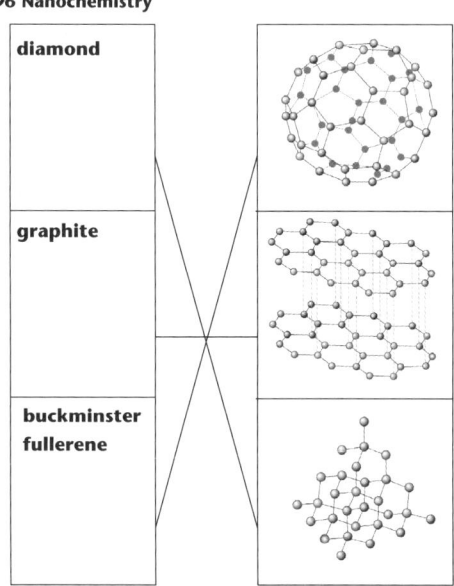

b
	diamond	graphite
use	cutting tools/jewellery	electrodes/pencil lead/lubricant
reason	very hard/lustrous and colourless	conduct electricity/ high melting point

c
diamond	graphite
Does not conduct electricity because it has no free electrons	Slippery because layers of carbon atoms are weakly held together and can slide easily over each other

2 a Allotropes
b In diamond, each atom is held by covalent bonds to four other atoms, tetrahedrally, which are bonded further in different directions; this is called a giant structure; so many strong covalent bonds make the diamond hard; the bonding results in no free electrons, so it does not conduct electricity

3 a Semi-conductors in electrical circuits; industrial catalysts; reinforcement of graphite in tennis rackets *(Any 2)*
b There is a very large surface area available

Page 97 How pure is our water?
1 a Sedimentation; filtration; chlorination
b Sedimentation: chemicals are added to make solid particles/bacteria settle out/larger bits drop to the bottom; filtration: a layer of sand on gravel filters out the remaining fine particles; some types of sand filter also remove microbes; chlorination: chlorine is added to kill microbes
c i Distillation is used to remove the dissolved substances
ii Distillation uses huge amounts of energy and is very expensive; it is only used when there is no fresh water

2 a Clean water saves more lives than medicines; that is why, after disasters and in developing countries, relief organisations concentrate on providing clean water supplies
b All water is recycled around the planet; if there is not enough rain in the winter, reservoirs do not fill up properly for the rest of the year

3 a Lead nitrate + potassium chloride → lead chloride + potassium nitrate
b Silver nitrate + potassium bromide → silver bromide + potassium nitrate
c $AgNO_{3(aq)} + KBr_{(aq)} \rightarrow AgBr_{(s)} + NaNO_{3(aq)}$

C5 How much?
Page 99 Moles and empirical formulae
1 a The relative atomic mass of an element is the average mass of an atom compared to the mass of 1/12th of an atom of carbon-12
b 132
c 1.62 g (2.5 is (125 ÷ 100) × 2; (81 ÷ 100) × 2 is 1.62)
d 0.02 moles/(2.5 ÷ 125)

2 a 2.3 g
b 17 g *(3 marks)* or $NaOH + HNO_3 \rightarrow NaNO_3 + H_2O$, 0.2 moles NaOH will give 0.2 moles $NaNO_3$ *(1)*; 1 mole $NaNO_3$ is 85 g *(1)*; 0.2 moles is 17 g *(1)*

3 a The simplest whole number ratio of each type of atom in a compound
b CH_2O
c CH_2O

C	H	O
7.2	1.2	9.6
÷ 12	÷ 1	÷ 16
0.6	1.2	0.6
÷ 0.6	÷ 0.6	÷ 0.6
1	2	1

Page 100 Electrolysis
1 a Hydrogen and oxygen
b Electrolysis is the decomposition of a liquid by passing an electric current through it, where the negative ions are attracted to the positive electrode and the positive ions are attracted to the negative electrode
c i Hydrogen ions will be discharged at the cathode in preference to potassium ions

ii
at the cathode	at the anode
H^+, K^+ attracted; $2H^+ + 2e^- \rightarrow H_2$	NO_3^-, OH^- attracted; $4OH^- - 4e^- \rightarrow 2H_2O + O_2$

2 a i It increases **ii** It decreases
b Both
c 0.18 g: 1.5 × 60 × 60 = 5400 s; 5400 × 0.1 = 540 C; 270 C is half of 540 C

3 a
molten electrolyte	at the cathode	at the anode
Al_2O_3	$Al^{3+} + 3e^- \rightarrow Al$	$2O^{2-} - 4e^- \rightarrow O_2$
$PbBr_2$	$Pb^{2+} + 2e^- \rightarrow Pb$	$2Br^- - 2e^- \rightarrow Br_2$

b So that the ions are free to move

Page 101 Quantitative analysis
1 a 10 g
b 50%
c The fibre content is the same
d 2.0 g salt

2 a 0.75 dm^3
b 20 cm^3
c i 1 mol/dm^3
ii 4 g/dm^3
iii 0.02 mol/dm^3 *(2)* or 40 g in 1000 cm3 is 1 mol/dm^3 *(1)*
0.8g in1000 cm^3 is 0.02 mol/dm^3 *(1)*

3 a More crowded particles
b She would add 90 cm^3 water and add 10 cm^3 of acid solution

Page 102 Titrations
1 a **A** pH is high/alkali
B Excess acid/pH is below 7
b 24 cm^3
c 1.6
d Acid and alkali neutralise making a salt and water only
e Top line starts at pH 13.5, vertical drop is at 23.5 cm^3 of acid; curves off at the bottom at pH 1.4

2 a To get consistent results
b 0.111 mol/dm^3: Number of moles of HCl used = concentration × volume, 0.12 × 23.1/1000, which is 2.772/1000 moles; since 1 mole of HCl neutralises 1 mole of NaOH, 2.772 ÷ 1000 moles HCl neutralises 0.002772 moles NaOH; 25 cm^3 NaOH contains 0.002772 moles, so 1000 cm^3 contains 0.002772 × 1000 ÷ 25 mol/dm^3

3 a Litmus gives a sudden change in colour, giving a sharp end point. Mixed indicators give a continuous colour change
b Litmus gives a sharp colour change at the end point of a titration, so it is easy to spot. Universal indicator give a continuous colour change

Page 103 Gas volumes
1 a He puts a flask on the balance and adds marble chips, he puts a loose plug of cotton wool in the mouth of the flask; he records the mass; he then adds a fixed volume of dilute acid and records the mass of the contents every 30 s, the mass will decrease as gas is given off
b She adds magnesium to a flask, she connects the syringe to the bung that will fit the flask; she then adds a fixed volume of dilute acid and quickly puts the bung in the mouth of the flask; she records the volume of gas produced every 10 s as the gas is given off

WORKBOOK ANSWERS

2 a i 22 cm³
  ii 56–60 s. The magnesium was all used up
  iii 16 cm³
  iv 12 ÷ 5 cm³/s = 2.4 cm³/s
 b i Starts at 0 less steep than line 1 (to the right of line 1), levels out at volume 11 cm³
  ii She only used half the mass of Mg
 c 0.12 dm³ She used 0.12 ÷ 24 moles Mg; 0.005 moles H_2 would be collected; 0.005 moles occupies 24 × 0.005 dm³

Page 104 Equilibria
1 a The rate of the forward reaction equals the rate of the backward reaction
 b To the right
 c The reaction must be a closed system; the forward rate will begin to decrease; the backward rate will begin to increase
2 a i 40%
  ii Increases
  iii Decreases
 b Moves to the right
 c There are 4 moles of reactants which react to make 2 moles of products. Increasing the pressure will help the reaction which reduces the number of moles
3 a i Burning sulfur in oxygen
  ii $S + O_2 \rightarrow SO_2$; $2SO_2 + O_2 \rightleftharpoons 2SO_3$; $SO_3 + H_2O \rightarrow H_2SO_4$
 b **450 °C**, a compromise temperature. The forward reaction is exothermic, so high temperatures reduce the yield, but the temperature needs to be high enough to increase the rate of reaction; **Atmospheric pressure**. A compromise. High pressure increases the yield, however, the equilibrium lies to the right, so the cost of stronger equipment is not worth it; **V_2O_5 catalyst**, does not affect the position of the equilibrium but does make the reaction go faster, so more is produced every second

Page 105 Strong and weak acids
1 a H^+
 b B ionises fully so that all the H^+ ions are available
 c Reversible reaction
 d $HCl \rightarrow H^+ + Cl^-$
 e $CH_3COOH \rightleftharpoons CH_3COO^- + H^+$
 f Strength: the degree to which the acid molecules ionise; Concentration: the number of moles of acid in 1 dm³ solution
 g The pH of a weak acid is higher than the pH of a strong acid as there are less available H^+ ions (and the scale is a negative logarithmic scale)
2 Ethanoic acid is a weaker acid than HCl, as it does not ionise fully in solution; so it produces less available H^+ ions than HCl; so there is a lower collision frequency of H^+ ions with Mg so a slower reaction
3 a There are fewer H^+ ions that can move
 b H^+ ions are positive so attracted to the negative electrode

Page 106 Ionic equations
1 a Ions are free to move
 b Silver nitrate and sodium bromide, react in a precipitation reaction to form silver bromide through collision of ions in solution; this is a very rapid reaction as the ions are completely free to move from the start
2 a Silver nitrate + sodium bromide → silver bromide + sodium nitrate
 b Silver nitrate + sodium chloride → silver chloride + sodium nitrate
 c Potassium sulphate + barium chloride → potassium chloride + barium sulphate
 d $Ag^+ + NO_3^- + K^+ + I^- \rightarrow AgI + K^+ + NO_3^-$
 e $Pb^{2+}_{(aq)} + I^-_{(aq)} \rightarrow PbI_{2(s)}$
3 a Add the solutions together; filter the precipitate; wash the precipitate with distilled water; dry the precipitate
 b i $Pb^{2+}_{(aq)} + 2NO_3^-_{(aq)} + 2K^+_{(aq)} + 2I^-_{(aq)} \rightarrow 2K^+_{(aq)} + 2NO_3^-_{(aq)} + PbI_{2(s)}$
  ii The spectator ions do not form part of the precipitate but remain in solution. The $K^+_{(aq)}$ and $NO_3^-_{(aq)}$ are in the solution as reactants and products and are, therefore, spectators

C6 Chemistry out there
Page 108 Energy transfers – fuel cells
1 a Exothermic
 b [energy profile diagram: reactants high, products low, arrow showing energy given out in reaction; axes: energy inside chemicals vs course of reaction]
2 a i Hydrogen + oxygen → water
  ii When hydrogen reacts with oxygen the chemical energy is converted directly into electrical energy, a potential difference is created
  iii $2H_2 + O_2 \rightarrow 2H_2O$
 b i Electrons are released as H_2 gas becomes H^+ ions
  ii Electrons are taken in as water and oxygen become hydroxyl ions OH^-
  iii Where electrons are lost and gained/where **red**uction and **ox**idation take place at the same time
3 a They are efficient – they waste very little energy; they are lighter than normal batteries; they do not need time out to be recharged; the water produced is used by the astronauts (Any 3)
 b The laws on pollution from carbon emissions are being tightened and there are no carbon emissions from the use of fuel cells; they are more efficient as no energy is lost as heat energy
 c Hydrogen gas is difficult to store and not yet readily available
 d More efficient; fewer stages; there is direct energy transfer; less pollution (water given off) (Any 2)

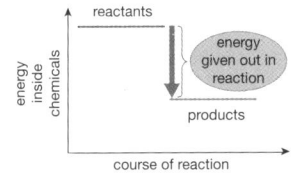

Page 109 Redox reactions
1 a Iron + water + oxygen → hydrated iron(III) oxide
 b Where electrons are lost and gained at the same time; where **red**uction and **ox**idation take place at the same time
2 a Paint/covering with grease/oil
 b i Zinc acts as a barrier and also corrodes instead of ion (sacrificial)
  ii A metal such as magnesium corrodes instead of iron as it loses electrons in preference to iron
  iii Acts as a barrier, but if scratched but will not act sacrificially

3 a Zinc metal is less reactive than magnesium so does not displace it from solution
 b Zinc + iron sulfate → zinc sulfate + iron
 c $Mg + ZnSO_4 \rightarrow MgSO_4 + Zn$
 d $Zn + Fe^{2+} \rightarrow Zn^{2+} + Fe$
4 a **Red**uction and **ox**idation occur at the same time
 b

	electrons lost or gained?	oxidised or reduced?
Fe	lost	oxidation
Fe^{2+}	gained	reduction

Page 110 Alcohols
1 a Glucose → carbon dioxide + ethanol
 b i Temperature too low, yeast inactive; temperature too high, enzyme in yeast is denatured
  ii Prevents formation of ethanoic acid
 c Distillation
 d i C_2H_5OH
  ii [structural formula of propan-1-ol: H-C(H)(H)-C(H)(H)-C(H)(H)-O-H]
  iii [structural formula of pentan-1-ol: H-C(H)(H)-C(H)(H)-C(H)(H)-C(H)(H)-C(H)(H)-O-H]
  iv $C_nH_{2n+1}OH$
2 a i Phosphoric acid **ii** Heated, catalyst
 b i Ethene + water → ethanol
  ii $C_2H_4 + H_2O \rightarrow C_2H_5OH$
3 a Aluminium oxide (by the bung)
 b Ethanol → ethene + water
 c $C_2H_5OH \rightarrow C_2H_4 + H_2O$

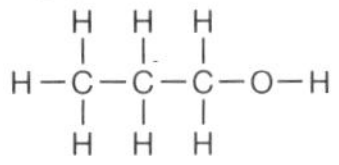

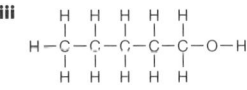

Page 111 Chemistry of sodium chloride (NaCl)
1 a Cheshire **b** Subsidence
2 a i Inert **ii** Anode **iii** Hydrogen
  iv Because Na^+ ions are not discharged they combine with OH^- ions to form NaOH
 b i $2Cl^- - 2e^- \rightarrow Cl_2$ **ii** $2H^+ + 2e^- \rightarrow H_2$
 c Oxygen
3 a $Na^+ + e^- \rightarrow Na$ **b** $2Cl^- - 2e^- \rightarrow Cl_2$
4 Chlorine; sodium hydroxide

Page 112 Depletion of the ozone layer
1 They deplete the ozone layer and as they are so stable they are only slowly removed and last for years in the upper atmosphere; the ozone layer protects the Earth from excessive UV light
2 a Cl **b** Free radical
 c They are very stable and are only slowly removed, so last for years in the upper atmosphere
 d Unevenly as ions; evenly as highly reactive free radicals
3 a Increased levels of UV light
 b $Cl\cdot + O_3 \rightarrow OCl\cdot + O_2$ $OCl\cdot + O_3 \rightarrow Cl\cdot + 2O_2$
4 a Alkanes/HFCs
 b 82%
 c The chain reaction set up terminates when 2 Cl free radicals meet; this takes some time

Page 113 Hardness of water
1 a Makes it slightly acidic
 b Calcium carbonate + water + carbon dioxide → calcium hydrogencarbonate
 c $CaCO_3 + H_2O + CO_2 \rightarrow Ca(HCO_3)_2$
2 a Ion-exchange resins are in a column. The water flows over solid resin which has sodium ions on it; the resin traps the calcium and magnesium ions on to it, taking these ions out of the water exchanging them for the sodium ions
 b After use the column has calcium and magnesium ions on it; they need to be removed and replaced by sodium ions, so NaCl is put into the softener
3 a Calcium hydrogencarbonate
 b $CaCO_3 + H^+ \rightarrow Ca^{2+} + H_2O + CO_2$
4 a All of the other samples lather well/better with soap after boiling; they were either not hard or had temporary hardness
 b Use the same volume of water; use the same volume of soap

Page 114 Natural fats and oils
1 a Esters
 b i All the carbon–carbon bonds are single **ii** Double bond circled
 c i Shake bromine water with a sample of each hydrocarbon
  ii With M, the brown colour would stay; with N, the bromine would decolourise
 d The saturated molecules do not react with the bromine molecules; the double bond in the unsaturated molecule reacts with the bromine making a di-bromo compound; the colour is therefore lost
2 a Saponification
 b Fat + sodium hydroxide → soap + glycerol
 c Hydrolysis reaction
3 People whose diet is rich in unsaturated oils usually have lower levels of cholesterol; high levels of cholesterol have been linked with an increase of risk of heart disease

Page 115 Analgesics
1 a $C_9H_8O_4$
 b i A benzene ring **ii** N **iii** Two COOH groups
 c i It is faster acting and has fewer side effects
  ii It is reacted with a base, such as sodium hydroxide, to make the COO^- ion which is soluble in water
2 a It is an externally administered substance that modifies or affects chemical reactions in the body
 b Nothing else in the drug must be allowed to affect the body and cause other problems
 c Salicylic acid and ethanoic anhydride are heated together for a day; they are cooled slowly, so they form large crystals; the crystals are filtered off; the crystals are purified

128